Diss. ETH No. 18347

Adaptive Quadrature Re-Revisited

A dissertation submitted to the
SWISS FEDERAL INSTITUTE OF TECHNOLOGY
ZÜRICH

for the degree of
DOCTOR OF SCIENCES

presented by
Pedro Gonnet

Dipl. Inf. Ing., ETH Zürich
born on January 6th, 1978
citizen of Zürich, Switzerland

Accepted on the recommendation of
Prof. Dr. W. Gander, examiner
Prof. Dr. J. Waldvogel, co-examiner
Prof. Dr. D. Laurie, co-examiner

2009

ISBN 978-0-557-08761-7

Acknowledgments

First and foremost I would like to thank my thesis advisor, Prof. Walter Gander who gave me the opportunity to switch to his group and offered me an interesting topic. I also thank him for introducing me to his many friends and acquaintances in what is, in my opinion, more of a *family* than a field.

From this family and the Department of Computer Science at large, I would also like to thank Oscar Chinellato, Prof. Jörg Waldvogel, Michael Bergdorf, Urs Zimmerli, Marco Wolf, Martin Müller, Marcus Wittberger, Cyril Flaig, Dirk Zimmer and Prof. François Cellier who all contributed in some part, through discussions or support, to the completion of this thesis.

I would also like to thank Prof. Milovanovic, Prof. Bojanov, Prof. Nikolov and Aleksandar Cvetkovic, as well as the many participants of the SCOPES meetings in Serbia, Bulgaria and here in Zurich for giving me a chance to present my work and learn from their questions and their criticisms.

Many thanks also go to E. H. A. Venter, F. J. Smith, E. de Doncker, P. Davis, T. O. Espelid and R. Jaffe for their help in retrieving and understanding some of the older or less accessible publications referenced in this thesis.

Many thanks go to my family: My mother, Marta, my younger brothers Julio and Ignacio for their support. Very special thanks go to my older sister Ariana, who took the time to proofread this dissertation. I thank my father, Gaston, for his support, advice and long conversations over coffee regarding computing, research, science, politics and life.

Most importantly, I would like thank Joanna, whom I love very much,

i

for her patience and understanding, for standing by me and for pushing me forward.

Pedro Gonnet

Zurich, December 2008

Abstract

The goal of this thesis is to explore methods to improve on the reliability of adaptive quadrature algorithms. Although adaptive quadrature has been studied for several years and many different algorithms and techniques have been published, most algorithms fail for a number of relatively simple integrands.

The most important publications and algorithms on adaptive quadrature are reviewed. Since they differ principally in their error estimates, these are tested and analyzed in detail and their similarities and differences are discussed.

In order to construct a more reliable error estimate, we start with a more complete and explicit representation of the integrand. To this extent, we **explicitly construct the interpolatory polynomial over the nodes of a quadrature rule**, which is constructed implicitly when the quadrature rule is evaluated.

The interpolatory polynomial is represented as a linear combination of orthogonal polynomials satisfying a three-term recurrence relation. Two new algorithms are presented for the efficient and numerically stable **construction, update and downdate** of such interpolations.

This representation can be easily **transformed** to any interval, thus allowing for its re-use inside the sub-intervals. The interpolation updates can be used to increase the degree of a quadrature rule by adding more nodes and the downdates can be used to remove nodes for which the integrand is not defined.

Based on this representation, **two new error estimates** based on the L_2-norm of the difference between the interpolant and the integrand (as opposed to using the integral of the difference between the interpolant and the integrand) are derived and shown to be more reliable than previously published error estimates. The space of possible integrands for which the new estimators will fail is shown to be smaller than that of previous error estimators.

A mechanism for the **detection of divergent integrals**, which may cause the algorithm to recurse infinitely and are usually caught by artificially limiting the number of function evaluations, is also included and tested.

These new improvements are implemented in **three new quadrature routines** and tested against several other commonly used routines. The results show that the new algorithms are **all more reliable** than previously published algorithms.

Zusammenfassung

Das Ziel dieser Dissertation ist es, Verfahren zu untersuchen, um die Zuverlässigkeit von Algorithmen zur adaptiven Quadratur zu verbessern. Obwohl an der adaptiven Quadratur seit mehreren Jahren geforscht wird und eine Vielzahl von Algorithmen und Techniken schon publiziert wurden, versagen die gängigen Verfahren für zum Teil sehr einfache Integranden.

Die wichtigsten Publikationen und Algorithmen werden hier besprochen. Da sie sich hauptsächlich in der Art der Fehlerschätzung unterscheiden, werden die Verfahren zur Fehlerschätzung im Detail untersucht und getestet. Die Gemeinsamkeiten und Unterschiede zwischen den einzelnen Verfahren werden diskutiert.

Um einen genaueren Fehlerschätzer zu konstruieren, wird zuerst eine vollständige und explizite Darstellung des Integranden angefangen. Um diese zu erhalten, wird das Interpolationspolynom über die Knoten der Quadraturregel explizit konstruiert, welches sonst von dieser nur implizit berechnet wird.

Das Interpolationspolynom wird als Linearkombination einer orthogonalen Polynomialbasis dargestellt. Die Polynome dieser Basis erfüllen eine dreigliedrige Rekursionsformel. Zwei neue Algorithmen zur Berechnung, zur Erweiterung und zur Reduktion einer solchen Interpolation werden vorgestellt und getestet.

Diese Darstellung kann später auf ein beliebiges Intervall transformiert werden, was die Wiederverwertung in den Subintervallen ermöglicht. Die Interpolations-Erweiterung kann dazu benutzt werden, den Grad einer Quadraturformel zu erhöhen, indem zusätzliche Punkte zur Interpolation hinzugefügt werden und die Reduktion kann benutzt werden, um Punkte zu entfernen an denen der Integrand nicht definiert ist.

Gestützt auf diese Darstellung werden zwei neue Fehlerschätzer konstruiert. Es wird gezeigt, dass die neuen Fehlerschätzer zuverlässiger als die bisher publizierten Verfahren funktionieren und dass der Raum aller Funktionen, für die sie nicht funktionieren, stärker eingeschränkt ist als bei anderen Verfahren.

Ein neuer Mechanismus zur Erkennung divergenter Integrale, die bei anderen Verfahren zu einer unendlichen Rekursion führen können und durch willkürliche Rekursionsbeschränkungen abgefangen werden, wird ebenfalls vorgestellt und getestet.

Diese Neuerungen werden in drei neue Quadraturverfahren implementiert und gegen mehrere etablierte Verfahren getestet. Die Resultate zeigen, dass die drei neue Algorithmen zuverlässiger sind als die anderen im Test.

Table of Contents

Chapter 1

Introduction

1.1 Goals

Since the publication of the first adaptive quadrature routines in 1962 [55, 68], much work has been done and published on the subject, introducing new quadrature rules, new termination criteria, new error estimates, new data types and new extrapolation algorithms.

With these improvements, adaptive quadrature has advanced both in terms of reliability and efficiency, yet for most algorithms it is still relatively simple to find integrands for which they will *fail*. These integrands are not just pathological cases that would rarely occur in normal applications, but include relatively simple functions which *do* appear in every day use.

The goal of this thesis is to explore new methods in adaptive quadrature to make the process *more reliable* as a whole. This should be done, if possible, without aversely impacting its *efficiency*.

To this end, we will focus on computing better *error estimates* by using an *explicit representation of the integrand*. In doing so, we will also be able to *re-use* more information over several levels of recursion and be able to treat singularities, non-numerical function values, divergent integrals and other problems in a more efficient way.

The resulting algorithms should be able to integrate a larger set of integrands reliably with little or no performance penalty as compared to current "state-of-the-art" algorithms.

1.2 Notation, Definitions and Fundamental Concepts

In the following, we will use the notation

$$Q_n^{(m)}[a, b] \approx \int_a^b f(x) \, \mathrm{d}x$$

For a quadrature rule Q of *degree* n and *multiplicity* m over the interval $[a, b]$.

In several publications the notations $Q_n^{(m)}[a, b]f$ or $Q_n^{(m)}[a, b](f(x))$ are used to specify which function the quadrature rule is to be applied to. Since in most cases it is usually clear what integrand is intended, we will use the latter notation only when it is not obvious, from the context, to which function the quadrature rule is to be applied.

We will use the term **degree** to denote the *algebraic degree of precision* of a quadrature rule. A quadrature rule is of degree n when it integrates all polynomials of degree $\leq n$ exactly, but not all polynomials of degree $n + 1$. This is synonymous with the *precise degree of exactness* as defined by Gautschi in [39] or the *degree of accuracy* as defined by Krommer and Überhuber in [53].

We will use the term **multiplicity** to denote the number of *sub-intervals* or *panels* of equal size in $[a, b]$ over which the quadrature rule is applied. Hence,

$$Q_n^{(m)}[a, b] = \sum_{k=1}^m Q_n[a + (k-1)h, a + kh], \quad h = \frac{b-a}{m}. \tag{1.1}$$

In [12], $Q_n^{(m)}[a, b]$ is referred to as a *compound* or *composite* quadrature rule.

The quadrature rule itself is evaluated as the linear combination of the integrand evaluated at the **nodes** $x_i \in [-1, 1]$, $i = 0 \ldots n$

$$Q_n[a, b] = \frac{b - a}{2} \sum_{i=0}^{n} w_i f \left(\frac{a + b}{2} + \frac{b - a}{2} x_i \right)$$

where the w_i are the weights of the quadrature rule Q_n. For notational simplicity, we have assumed that a quadrature rule of degree n has $n + 1$ nodes. This is of course not always true: The degree of interpolatory quadrature rules over a symmetric set of an odd number of nodes is equal to the number of nodes, and for Gauss quadratures, Gauss-Lobatto quadratures and their Kronrod extensions, the degree is much larger than the number of nodes. The opposite may of course also be true: a quadrature rule need not be interpolatory and the degree may therefore be smaller than the number of nodes used by the rule (*e.g.* see Section 2.19).

Given a quadrature rule $Q_n^{(m)}[a, b]$ of degree n and multiplicity m applied to an integrand $f(x)$ which has $n + 1$ continuous derivatives in $[a, b]$ and for which the $n + 1$st derivative does not change sign in $[a, b]$, we model the error of the quadrature rule as

$$Q_n^{(m)}[a, b] - \int_a^b f(x)\,\mathrm{d}x = \kappa_n \left(\frac{b - a}{m} \right)^{n+2} f^{(n+1)}(\xi), \quad \xi \in [a, b]. \quad (1.2)$$

where the factor κ_n is the integral of the *Peano kernel* of the quadrature rule $Q_n^{(m)}$ over the interval $[a, b]$ (see [12], Section 4.3). If $f^{(n+1)}(x)$ is more or less constant and does not change sign for $x \in [a, b]$, the integrand is usually said to be "*sufficiently smooth*".

Equation 1.2 can be derived using the Taylor expansion of $f(x)$ around $x = a$:

$$f(a + x) =$$

$$f(a) + f'(a)x + \cdots + \frac{f^{(n)}(a)}{n!}x^n + \frac{f^{(n+1)}(\xi)}{(n+1)!}x^{n+1}, \quad \xi \in [a, b]. \quad (1.3)$$

Integrating both sides of Equation 1.3 from 0 to $b - a$, we obtain

$$\int_0^{b-a} f(a + x)\, \mathrm{d}x$$

$$= f(a) \int_0^{b-a} \mathrm{d}x + f'(a) \int_0^{b-a} x\, \mathrm{d}x + \ldots$$

$$+ \frac{f^{(n)}(a)}{n!} \int_0^{b-a} x^n\, \mathrm{d}x + \frac{f^{(n+1)}(\xi)}{(n+1)!} \int_0^{b-a} x^{n+1}\, \mathrm{d}x, \quad \xi \in [a, b]$$

$$= f(a)(b - a) + f'(a) \frac{(b - a)^2}{2} + \ldots$$

$$+ \frac{f^{(n)}}{n!} \frac{(b - a)^{n+1}}{n + 1} + \frac{f^{(n+1)}(\xi)}{(n+1)!} \frac{(b - a)^{n+2}}{n + 2}\, \mathrm{d}x, \quad \xi \in [a, b]. \quad (1.4)$$

Note that se can only extract $f^{(n+1)}(\xi)$ from the last term using the mean value theorem if $f^{(n+1)}(x)$ does not change sign in $x \in [a, b]$. Applying the *basic* (*i.e.* non-compound) quadrature rule $Q_n[0, b - a]$ on both sides of Equation 1.3 we obtain:

$$Q_n[0, b - a](f(a + x))$$

$$= f(a) Q_n[0, b - a](1) + f'(a) Q_n[0, b - a](x) + \ldots$$

$$+ \frac{f^{(n)}(a)}{n!} Q_n[0, b - a](x^n) + \frac{f^{(n)}(\xi)}{(n+1)!} Q_n[0, b - a](x^{n+1}).$$

$$(1.5)$$

Subtracting Equation 1.4 from Equation 1.5 we then obtain

$$Q_n[0, b - a](f(a + x)) - \int_0^{b-a} f(a + x)\, \mathrm{d}x$$

$$= \sum_{i=0}^{n+1} \frac{f^{(i)}(a)}{i!} Q_n[0, b - a](x^i) - \sum_{i=0}^{n+1} \frac{f^{(i)}(a)}{i!} \frac{(b - a)^{i+1}}{i + 1}$$

$$= \frac{f^{(n+1)}(\xi)}{(n+1)!} \left(Q_n[0, b - a](x^{n+1}) - \frac{(b - a)^{n+2}}{n + 2} \right)$$

$$= \kappa_n (b - a)^{n+2} f^{(n+1)}(\xi). \quad (1.6)$$

Since Q_n is exact for all polynomials of degree up to n, the first $n + 1$ terms in the right-hand sides of Equation 1.4 and Equation 1.5 cancel each other out. The remaining, final term of Equation 1.5, when inserted in Equation 1.1, results in a multiple of $(b-a)^{n+2}$:

$$
\begin{aligned}
Q_n[0, b-a](x^{n+1}) &= \frac{b-a}{2} \sum_{i=0}^{n} w_i \left(\frac{b-a}{2} x_i \right)^{n+1} \\
&= \left(\frac{b-a}{2} \right)^{n+2} \sum_{i=0}^{n} w_i x_i^{n+1} \\
&= (b-a)^{n+2} C
\end{aligned}
$$

where C is a constant which is independent of the interval of integration. This easily extends to the compound rule $Q_n^{(m)}$, the error of which can be computed as

$$
\begin{aligned}
& Q_n^{(m)}[a, b] - \int_a^b f(x)\, \mathrm{d}x \\
&= \sum_{i=1}^{m} \left[Q_n[a+(i-1)h, a+ih] - \int_{a+(i-1)h}^{a+h} f(x)\, \mathrm{d}x \right], \quad h = \frac{b-a}{m} \\
&= \sum_{i=1}^{m} \kappa_i h^{n+2} f^{(n+1)}(\xi_i), \quad \xi_i \in [a+(i-1)h, a+ih] \\
&= \kappa h^{n+2} f^{(n+1)}(\xi), \quad \xi \in [a, b] \\
&= \kappa \left(\frac{b-a}{m} \right)^{n+2} f^{(n+1)}(\xi)
\end{aligned}
$$

which is the error in Equation 1.2.

If a quadrature rule is of degree n, then its *order of accuracy* as defined by Skeel and Keiper in [93], to which we will simply refer to as its *order*, is the exponent of the interval widht $b-a$ in the error expression. For a quadrature rule of degree n, the order is $n+2$.

To avoid confusion regarding the nature of the degree vs. the number of nodes, we will use the following symbols to describe some special quadrature rules:

- $\mathsf{S}^{(m)}[a, b]$: compound Simpson rule of degree 3 over m sub-intervals,

- $\mathsf{NC}_n^{(m)}[a, b]$: compound Newton-Cotes rule of degree n over m sub-intervals (usually n will be odd and the number of nodes will also be n),

- $\mathsf{CC}_n[a, b]$: Clenshaw-Curtis quadrature over n *nodes* (note that here the subscript denotes the number of *nodes*, not the *degree*),

- $\mathsf{G}_n[a, b]$: Gauss quadrature over n *nodes* (note that here the subscript denotes the number of *nodes*, not the *degree*),

- $\mathsf{GL}_n[a, b]$: Gauss-Lobatto quadrature over n *nodes* (note that here the subscript denotes the number of *nodes*, not the *degree*),

- $\mathsf{K}_n[a, b]$: Kronrod extension over n *nodes* (note that here the subscript denotes the number of *nodes*, not the *degree*).

During adaptive integration, the interval $[a, b]$ is divided into a number of *sub-intervals* $[a_k, b_k]$ such that

$$\bigcup_k [a_k, b_k] = [a, b],$$

over which the quadratures are computed. The sum of the *local* quadratures is an approximation to the *global* integral:

$$\sum_k Q_n^{(m)}[a_k, b_k] \approx \int_a^b f(x) \, \mathrm{d}x.$$

The sub-division occurs either recursively or explicitly. Most algorithms follow either the *recursive* scheme in Algorithm 1 or the *non-recursive* scheme in Algorithm 2.

In both cases, the decision whether to sub-divide an interval or not is made by first approximating the integral over the sub-interval (Line 1 in Algorithm 1 or Lines 7 and 8 in Algorithm 2) and then approximating the *integration error* over the sub-interval (Line 2 in Algorithm 1 or Lines 9 and 10 in Algorithm 2). If this *error estimate* is larger than the accepted tolerance (Line 3 in Algorithm 1) or the largest error of all sub-intervals (Line 5 in Algorithm 2), then that interval is sub-divided.

Algorithm 1 integrate (f, a, b, τ)

1: $Q_n^{(m)}[a, b] \approx \int_a^b f(x)\, dx$ (*approximate the integral*)

2: $\varepsilon \approx \left| Q_n - \int_a^b f(x)\, dx \right|$ (*approximate the integration error*)

3: **if** $\varepsilon < \tau$ **then**

4:     **return** Q_n (*accept the estimate*)

5: **else**

6:     $m \leftarrow (a + b)/2$

7:     **return** integrate(f, a, m, τ') + integrate(f, m, b, τ') (*recurse*)

8: **end if**

Algorithm 2 integrate (f, a, b, τ)

1: $I \leftarrow Q_n^{(m)}[a, b] \approx \int_a^b f(x)\, dx$ (*initial approximation of I*)

2: $\varepsilon \leftarrow \varepsilon_0 \approx \left| Q_n - \int_a^b f(x)\, dx \right|$ (*initial approximation the error*)

3: initialize H with interval $[a, b]$, integral $Q_n[a, b]$ and error ε_0

4: **while** $\varepsilon > \tau$ **do**

5:     $k \leftarrow$ index of interval with largest ε_k in H

6:     $m \leftarrow (a_k + b_k)/2$

7:     $I_{\text{left}} \leftarrow Q_n^{(m)}[a, m] \approx \int_{a_k}^m f(x)\, dx$ (*evaluate $Q_n^{(m)}$ on the left*)

8:     $I_{\text{right}} \leftarrow Q_n^{(m)}[m, b] \approx \int_m^{b_k} f(x)\, dx$ (*evaluate $Q_n^{(m)}$ on the right*)

9:     $\varepsilon_{\text{left}} \approx \left| Q_n[a_k, m] - \int_{a_k}^m f(x)\, dx \right|$ (*approximate ε on the left*)

10:    $\varepsilon_{\text{right}} \approx \left| Q_n[m, b_k] - \int_m^{b_k} f(x)\, dx \right|$ (*approximate ε on the right*)

11:    $I \leftarrow I - I_k + I_{\text{left}} + I_{\text{right}}$ (*update the global integral*)

12:    $\varepsilon \leftarrow \varepsilon - \varepsilon_k + \varepsilon_{\text{left}} + \varepsilon_{\text{right}}$ (*update the global error*)

13:    push interval $[a_k, m]$ with integral I_{left} and error $\varepsilon_{\text{left}}$ onto H

14:    push interval $[m, b_k]$ with integral I_{right} and error $\varepsilon_{\text{right}}$ onto H

15: **end while**

16: **return** I

In the following, we will distinguish between the *local* and *global* error in an adaptive quadrature routine. The *local error* ε_k of the kth interval $[a_k, b_k]$ is defined as

$$\varepsilon_k = \left| Q_n[a_k, b_k] - \int_{a_k}^{b_k} f(x)\,\mathrm{d}x \right| \tag{1.7}$$

and the *global error* ε is

$$\varepsilon = \left| \sum_k Q_n[a_k, b_k] - \int_a^b f(x)\,\mathrm{d}x \right|. \tag{1.8}$$

Since the local errors may be of different sign, the sum of the local errors forms an upper bound for the global error.

$$\varepsilon \leq \sum_k \varepsilon_k.$$

We further distinguish between the absolute errors (1.7) and (1.8), the *locally relative local error*

$$\varepsilon_k^{\mathrm{lrel}} = \left| \frac{Q_n[a_k, b_k] - \int_{a_k}^{b_k} f(x)\,\mathrm{d}x}{\int_{a_k}^{b_k} f(x)\,\mathrm{d}x} \right|, \tag{1.9}$$

and the *globally relative local error*

$$\varepsilon_k^{\mathrm{grel}} = \left| \frac{Q_n[a_k, b_k] - \int_{a_k}^{b_k} f(x)\,\mathrm{d}x}{\int_a^b f(x)\,\mathrm{d}x} \right|. \tag{1.10}$$

We also define the *global relative error* which is bounded by the sum of the globally relative local errors:

$$\varepsilon^{\mathrm{grel}} = \left| \frac{\sum_k Q_n[a_k, b_k] - \int_a^b f(x)\,\mathrm{d}x}{\int_a^b f(x)\,\mathrm{d}x} \right| \leq \sum_k \left| \frac{Q_n[a_k, b_k] - \int_{a_k}^{b_k} f(x)\,\mathrm{d}x}{\int_a^b f(x)\,\mathrm{d}x} \right|.$$

The sum of the *locally relative local errors*, however, form no such bound: If the integral in a sub-interval is larger than the global integral, then the local error therein will be smaller relative to the local integral than relative to the global integral and may cause the sum of locally relative local errors to be smaller than the global relative error.

1.3 Outline

In Chapter 2, a detailed history of the most significant advances in adaptive quadrature algorithms from the past 45 years is given.

In Chapter 3 different ways in which the integrand is represented in different algorithms are discussed, as well as how it may be explicitly represented using orthogonal polynomials. A new algorithm for the efficient and stable construction, update and downdate (*i.e.* the addition or removal of a node, respectively) of such interpolations is described, as well as a new algorithm for their construction. Both algorithms are tested against previous efforts in the field.

In Chapter 4 the different types of error estimates described and used to date are discussed and categorized. A new type of error estimate is proposed and described in two distinct variants. These two new estimates are compared to all of the error estimators discussed previously. The results are then analyzed and a further classification of the error estimators is suggested.

In Chapter 5 some common problems of adaptive quadrature algorithms are discussed and strategies for dealing with singularities and non-numerical function values, as well as divergent integrals, are presented.

In Chapter 6 some general design considerations are discussed and three new algorithms are presented. These algorithms are compared to other widely-used algorithms on several sets of test-functions and shown to be more reliable.

In Chapter 7 we summarize the results from the previous chapters and draw some general conclusions.

Appendices A and B contain the Matlab source of the different error estimators and quadrature algorithms as they were implemented and tested for this thesis.

Chapter 2

A Short History of Adaptive Quadrature

2.1 A Word on the "First" Adaptive Quadrature Algorithm

There seems to be some confusion as to who actually published the first adaptive quadrature algorithm. Davis and Rabinowitz [12] cite the works of Villars [100], Henriksson [45] and Kuncir (see Section 2.2), whereas other authors [24, 20, 22, 21, 5, 66] credit McKeeman (see Section 2.3).

Although no explicit attribution is given, Henriksson's algorithm seems to be an unmodified ALGOL-implementation of the algorithm described by Villars. The algorithm described by Villars is, in turn, as the author himself states, only a slight modification of a routine developed by Helen Morrin [72][1] in 1955. These three algorithms already contain a number of features, *e.g.* the use of an analytical scaling factor for the error and subtraction the error from the integral estimate (in effect using Romberg extrapolation), which were apparently forgotten and later re-discovered

---

[1]This reference is only given for the sake of completeness since the author was not able to obtain a copy of the original manuscript. The routine presented therein is, however, described in detail by Villars in [100].

by other authors.

The algorithms of Morrin, Villars and Henriksson, however, are more reminiscent of ODE-solvers. Following the general scheme in Algorithm 3, they integrate the function stepwise from left to right using Simpson's rule and adapting (doubling or halving) the step-size whenever an estimate converges or fails to do so. These algorithms effectively discard function evaluations and so lose information on the structure of the integrand (see Figure 2.1). We will therefore not consider them to be "genuine" adaptive integrators.

Algorithm 3 ODE-like integration algorithm

$h \leftarrow b - a$	(*initial step size*)
$I \leftarrow 0$	(*initial integral*)
while $a < b$ **do**	
$\quad Q_n^{(m)}[a, a+h] \approx \int_a^{a+h} f(x)\,\mathrm{d}x$	(*approximate the integral*)
$\quad \varepsilon \approx \left| Q_n^{(m)}[a, a+h] - \int_a^{a+h} f(x)\,\mathrm{d}x \right|$	(*approximate the error*)
$\quad$ **if** $\varepsilon < \tau$ **then**	
$\qquad I \leftarrow I + Q_n^{(m)}[a, b]$	(*increment the global integral*)
$\qquad a \leftarrow a + h$	(*move the left boundary*)
$\qquad h \leftarrow \min\{2h, b - a\}$	(*increase the step size*)
$\quad$ **else**	
$\qquad h \leftarrow h/2$	(*use a smaller step size*)
$\quad$ **end if**	
end while	
return I	(*return the accumulated integral*)

This is not the case with Kuncir's algorithm (see Section 2.2), which follows the recursive scheme presented in Algorithm 1, and seems to be the first published algorithm to do so.

Kuncir predates McKeeman by approximately six months. Interestingly enough, the very similar works of both Kuncir and McKeeman were both published in the same journal (Communications of the ACM) in the same year (1962) in different issues of the same volume (Volume 5), both edited by the same editor (J.H. Wegstein). This duplication of efforts does not seem to have been noticed at the time.

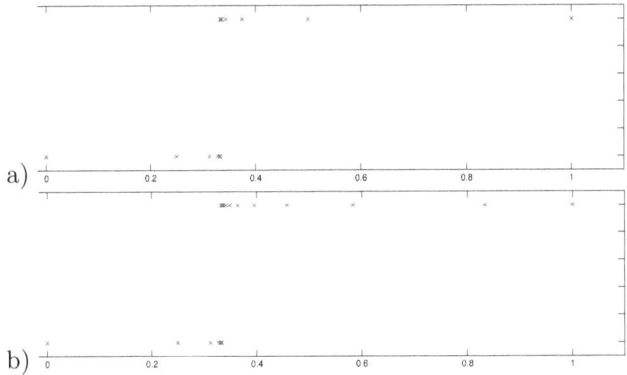

Figure 2.1: Boundaries of the intervals generated in (a) a recursive scheme using bisection and in (b) an ODE-like scheme halving or doubling the interval size depending on the success of integration in that interval as per Algorithm 3. If we subdivide up to an interval width of 10^{-5} around a difficulty at $x = 1/3$, the ODE-like scheme generates 25 intervals whereas the recursive scheme generates only 18.

2.2 Kuncir's Simpson Rule Integrator

In 1962, Kuncir [55] publishes the first adaptive quadrature routine following the scheme in Algorithm 1 and using Simpson's compound rule over two sub-intervals to approximate the integral

$$\int_a^b f(x)\,\mathrm{d}x \approx \mathsf{S}^{(2)}[a, b]$$

and the *locally relative* local error estimate

$$\varepsilon_k = \left| \mathsf{S}^{(1)}[a_k, b_k] - \mathsf{S}^{(2)}[a_k, b_k] \right| \frac{2^d}{\left| \mathsf{S}^{(2)}[a_k, b_k] \right|} \tag{2.1}$$

where d is the recursion depth of the kth interval $[a_k, b_k]$. The function values of the compound rule can be re-used to compute the basic rule $\mathsf{S}^{(1)}$ in the sub-intervals after bisection.

Replacing every evaluation of the integrand in the un-scaled error estimate (Equation 2.1) with an appropriate $f(a + h)$ and expanding it in a

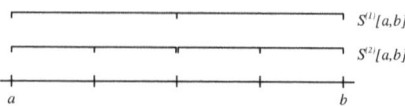

Figure 2.2: Kuncir's error estimate uses the difference between Simpson's rule applied once ($S^{(1)}[a, b]$) and Simpson's rule applied twice ($S^{(2)}[a, b]$) in the same interval.

Taylor expansion around a, as is done in [32], we obtain

$$S^{(1)}[a_k, b_k] - S^{(2)}[a_k, b_k] = \frac{(b_k - a_k)^5}{3072} f^{(4)}(\xi), \quad \xi \in [a_k, b_k], \qquad (2.2)$$

from which we see that the error is a function of the fourth derivative of the integrand. Inserting the Taylor expansion into the *actual* error gives a similar result:

$$S^{(2)}[a_k, b_k] - \int_{a_k}^{b_k} f(x)\, dx = \frac{(b_k - a_k)^5}{46\,080} f^{(4)}(\xi), \quad \xi \in [a_k, b_k]. \qquad (2.3)$$

If we assume that $f^{(4)}(x)$ is more or less constant for $x \in [a_k, b_k]$ and both Equation 2.2 and Equation 2.3 therefore have similar values for $f^{(4)}(\xi)$, then the error estimate is 15 times larger than the actual integration error. This factor of 15 might seem large, but in practice it is a good guard against bad estimates when $f^{(4)}(x)$ is *not* constant for $x \in [a_k, b_k]$.

The scaling factor 2^d, which is implemented implicitly by scaling the local tolerance (τ' in Line 7 of Algorithm 1) by a factor of $1/2$ upon recursion (*i.e.* $\tau' \leftarrow \tau/2$), was probably chosen to ensure that, after subdividing an interval, the sum of the interval error estimates can not exceed the requested tolerance. However, since a *locally relative* error estimate is used, this is only true if the integral over the sub-intervals is of approximately the same magnitude. If the integral over any sub-interval is larger in magnitude than the global integral, then its local error, while still below the locally relative tolerance, may exceed the global relative tolerance.

This can lead the algorithm to fail to converge for integrals which tend to zero locally. Consider, for example, the integration of $f(x) = x^{1.1}$ over the interval $[0, 1]$. The expansion of the error of the leftmost interval

$[0, h]$ results in:

$$
\begin{aligned}
\varepsilon &= \left| \mathsf{S}^{(1)}[0,h](x^{1.1}) - \mathsf{S}^{(2)}[0,h](x^{1.1}) \right| \frac{2^d}{\left| \mathsf{S}^{(2)}[0,h](x^{1.1}) \right|} \\
&= \left| \mathsf{S}^{(1)}[0,2^{-d}](x^{1.1}) - \mathsf{S}^{(2)}[0,2^{-d}](x^{1.1}) \right| \frac{2^d}{\left| \mathsf{S}^{(2)}[0,2^{-d}](x^{1.1}) \right|} \\
&\approx 0.0024 \times 2^d.
\end{aligned}
\tag{2.4}
$$

where the error estimate *grows exponentially* with each subdivision. If the error criterion is not met in the first step of the algorithm, it never will be met. Kuncir guards against this problem by using a user-supplied maximum recursion depth N. This prevents the algorithm from recursing indefinitely, yet does not guarantee that the requested tolerance will indeed have been achieved.

2.3 McKeeman's Adaptive Integration by Simpson's Rule

In the same year, McKeeman [68] publishes a similar recursive algorithm (following Algorithm 1, yet using trisection instead of bisection) using the compound Simpson's rule over three panels to approximate the integral

$$
\int_a^b f(x)\,\mathrm{d}x \approx \mathsf{S}^{(3)}[u,b]
$$

and the *globally relative* local error estimate

$$
\varepsilon_k = \left| \mathsf{S}^{(1)}[a_k,b_k] - \mathsf{S}^{(3)}[a_k,b_k] \right| \frac{3^d}{\hat{I}_d}
\tag{2.5}
$$

(see Figure 2.3) where $\hat{I}_d$ is an approximation to the integral of the absolute value of $f(x)$:

$$
\hat{I}_d \approx \int_a^b |f(x)|\,\mathrm{d}x.
$$

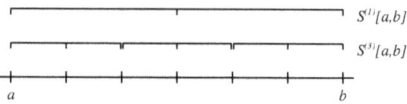

Figure 2.3: McKeeman's error estimate uses the difference between Simpson's rule applied once ($\mathsf{S}^{(1)}[a,b]$) and Simpson's rule applied three times ($\mathsf{S}^{(3)}[a,b]$) in the same interval.

This estimate is updated in every recursion step using

$$\hat{I}_d =$$

$$\hat{I}_{d-1} - \left| \mathsf{S}^{(1)}[a_k, b_k] \right| + \sum_{i=1}^{3} \left| \mathsf{S}^{(1)}[a_k + (i-1)h, a_k + ih] \right|,$$

$$h = \frac{b_k - a_k}{3}. \quad (2.6)$$

The main differences with respect to Kuncir's error estimator can be summed-up as follows:

1. The use of *trisection* as opposed to bisection,

2. The use of a *globally* relative local error estimate as opposed to a locally relative error estimate,

3. The error is computed relative to the global integral of the *absolute value* of $f(x)$ as opposed to the local integral of $f(x)$.

Replacing the evaluations of the integrand $f(a+h)$ by their Taylor expansions around a in Equation 2.5, as we did with Kuncir's error estimator (see Equation 2.2), we can see that

$$\left| \frac{\mathsf{S}^{(1)}[a,b] - \mathsf{S}^{(3)}[a,b]}{\mathsf{S}^{(3)}[a,b] - \int_a^b f(x)\,\mathrm{d}x} \right| \approx 80, \quad (2.7)$$

i.e. the error is overestimated by a factor of 80 for a sufficiently smooth integrand. This is significantly stricter than the factor of 15 in Kuncir's algorithm. Furthermore, although the scaling factor 3^d guarantees that the sum of the error estimates will not exceed the total tolerance, it is very strict in practice and will cause the algorithm to fail to converge around certain types of singularities.

Due to the *globally* relative error estimate, this algorithm has no difficulty integrating $f(x) = x^{1.1}$ (see Section 2.2, Equation 2.4), yet consider the case of integrating $f(x) = x^{-1/2}$ in $[\varepsilon_{mach}, 1]$: The function has a singularity at $f(0) = \infty$ (which is why we integrate from ε_{mach}) but its integral is bounded. Applying a similar analysis as for Kuncir's error estimate (Equation 2.4), we get, for the leftmost interval:

$$\begin{aligned} \varepsilon &= \left| S^{(1)}[\varepsilon_{mach}, 3^{-d}](x^{-1/2}) - S^{(3)}[\varepsilon_{mach}, 3^{-d}](x^{-1/2}) \right| 3^d \times 2 \\ &\approx 1.5 \times 10^7. \end{aligned} \tag{2.8}$$

The error estimate does not grow as it would using Kuncir's estimate ($\varepsilon \sim 2^d$) since the error is relative to the global integral, yet it still does not decrease with increasing recursion depth.

The same type of problem occurs with the integrand $f(x) = x^{1/2}$, which is smooth and has a bounded integral, yet also has a singularity in the first derivative at $x = 0$. Using the same analysis as above when integrating in $[0, 1]$ we obtain for the leftmost interval

$$\begin{aligned} \varepsilon &= \left| S^{(1)}[0, 3^{-d}](x^{1/2}) - S^{(3)}[0, 3^{-d}](x^{1/2}) \right| 3^d \times 2/3 \\ &\approx -0.725 \times 3^{d/2} \end{aligned} \tag{2.9}$$

in which the error grows *exponentially* with the increasing recursion depth. Similarly to Kuncir's algorithm, this type of problem is caught using a fixed maximum recursion depth of 7.

The use of a *globally relative* local error estimate is an important improvement. Besides forming a correct upper bound for the global error, it does not run into problems in sub-intervals where the integrand approaches 0, causing any *locally relative* error estimate to approach infinity.

Another uncommon yet beneficial feature is the use of the global integral of the *absolute* value of the function. This is a good guard against cancellation or *smearing* [44] when summing-up the integrals over the sub-intervals. Consider, for example, the computation of $\int_0^{2\pi} 10\,000\,sin(5x)\,dx$: the integral itself is 0, yet when summing the non-zero integrals in the sub-intervals, we can expect cancellation errors of up to $10\,000\varepsilon_{\text{mach}}$ or 10^{-12} for IEEE 754 double-precision machine arithmetic[2]. It would therefore make little sense to pursue a higher precision in the sub-intervals, which is what this estimate guards against. However, this error relative to $\hat{I}$ is generally *not* what the user wants or is willing to think about: he or she is usually only interested in the effective (and not the feasible) number of correct digits in the result.

2.4 McKeeman and Tesler's Adaptive Integrator

A year later, McKeeman and Tesler publish a non-recursive[3] [71] version of McKeeman's Simpson Integrator (see Section 2.3). This integrator is, in essence, almost identical to its predecessor, yet with one fundamental difference, which had been suggested by McKeeman himself as an improvement in [70], namely:

$$\varepsilon_k = \left| \mathsf{S}^{(1)}[a_k, b_k] - \mathsf{S}^{(3)}[a_k, b_k] \right| \frac{1.7^d}{\hat{I}} \qquad (2.10)$$

where $\hat{I}$ is computed as in Equation 2.6 yet stored and updated *globally*. After each trisection, the error is no longer scaled with a factor of three, but with $1.7 \approx \sqrt{3}$. This small, uncommented[4] change is actually a significant improvement to the algorithm.

This relaxation can be derived statistically. Let us assume that the error of a quadrature rule $Q[a, b]$ can be described in terms of a normal distribution around 0:

$$Q[a, b] - \int_a^b f(x)\,\mathrm{d}x = \mathcal{N}(0, (\kappa\tau)^2) \qquad (2.11)$$

---

[2]indeed, QUADPACK's `QAG`, for instance, achieves -1.4×10^{-11}, MATLAB's `quad` achieves -1.3×10^{-12}, both for a requested absolute tolerance of 10^{-15}

[3]The algorithm of McKeeman and Tesler [71] is non-recursive in the sense that an explicit stack is maintained, analogous to the one generated in memory during recursion, and not as in the scheme presented in Algorithm 2

[4]A similar factor is later used by McKeeman (see Section 2.5) with a comment that dividing the tolerance by the number of sub-intervals, *i.e.* scaling by m instead of $\sqrt{m}$, "*proves too strict in practice*".

where the variance $(\kappa\tau)^2$ is a function of the required tolerance τ with (hopefully) $\kappa < 1$. If we subdivide the interval into m equally sized sub-intervals and compute the quadrature with some adjusted tolerance τ', the error should therefore behave as

$$
\begin{aligned}
\sum_{k=1}^{m} Q[a_k, b_k] - \int_a^b f(x)\,\mathrm{d}x &= \sum_{i=1}^{m} \left[Q[a_k, b_k] - \int_{a_k}^{b_k} f(x)\,\mathrm{d}x \right] \\
&= \sum_{i=1}^{m} \mathcal{N}(0, (\kappa\tau')^2) \\
&= \mathcal{N}(0, m(\kappa\tau')^2)
\end{aligned}
\tag{2.12}
$$

i.e. the variance of the sum of the errors is the sum of the variances of the errors. If we now want the two estimates (Equation 2.11 and Equation 2.12) to have the same variance, we must equate

$$
m(\kappa\tau')^2 = (\kappa\tau)^2 \quad \rightarrow \quad \tau' = \tau m^{-\frac{1}{2}}
\tag{2.13}
$$

resulting in the scaling by $1.7 \approx \sqrt{3}$ used by McKeeman and Tesler. When applied to the problems in Equation 2.8 and Equation 2.9, this modification causes the error estimate in the former to converge and that of the latter to remain constant (as opposed to exponential growth). This simple idea has been, as we shall see, subsequently re-used by many authors.

Such a model was already introduced by Henrici in [43, Chapter 16] for the accumulation of rounding errors in numerical calculations. Whereas in such a case, the independence of the individual errors is a valid assumption, in the case of integration errors it is not[5].

2.5 McKeeman's Variable-Order Adaptive Integrator

Shortly after publishing the "non-recursive" recursive adaptive integrator, McKeeman publishes another recursive adaptive integrator [69] based

---

[5]Assume, for example, that the error behaves as in Equation 1.2 where $f^{(n+1)}(x)$ is of constant sign for $x \in [a, b]$, which is the usual assumption for a "sufficiently smooth" integrand.

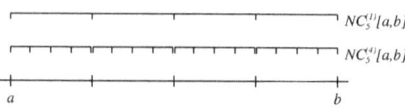

Figure 2.4: McKeeman's variable order error estimate uses the difference between a Newton-Cotes rule over n points applied once ($NC_n^{(1)}[a, b]$) and the same Newton-Cotes rule applied $n-1$ times ($NC_n^{(n-1)}[a, b]$) in the same interval (shown here for $n = 5$).

on Newton-Cotes rules over n points, where n is supplied by the user. In the same vein as the previous integrator, the integrand is computed using the compound Newton-Cotes rule over $n - 1$ panels

$$\int_a^b f(x)\, dx \approx NC_n^{(n-1)}[a, b]$$

and the error estimate is computed as

$$\varepsilon_k = \left| NC_n^{(1)}[a_k, b_k] - NC_n^{(n-1)}[a_k, b_k] \right| \frac{\sqrt{n}^d}{\hat{I}_d} \tag{2.14}$$

where $NC_n^{(m)}[a, b]$ is the Newton-Cotes rule over n points applied on m panels in the interval $[a, b]$ (see Figure 2.4). At every recursion level, the interval is subdivided into $n - 1$ panels, effectively re-using the nodes of the composite rule as the basic rule in each sub-interval.

This algorithm is a good example of the pitfalls of using empirical error estimates of increasing degree without considering how the error estimate behaves analytically. Replacing the evaluations of the integrand $f(a+h)$ by their Taylor expansions around a and inserting them into the expression

$$\left| \frac{NC_n^{(1)}[a, b] - NC_n^{(n-1)}[a, b]}{NC_n^{(n-1)}[a, b] - \int_a^b f(x)\, dx} \right|$$

we can see that for $n = 3$ (*i.e.* applying Simpson's rule), we overestimate the actual error by a factor of 15 for sufficiently smooth $f(x)$, as we had already observed in Kuncir's algorithm (see Section 2.2).

For $n = 4$, this factor grows to 80, as observed for McKeeman's first integrator (see Section 2.3). For $n = 5$ it is 4095 and for $n = 8$, the

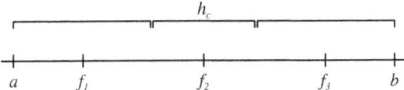

Figure 2.5: Gallaher's error estimate approximates the second derivative of $f(x)$ using divided differences over the three function evaluations f_1, f_2 and f_3.

maximum allowed in the algorithm, it is 5 764 800 (7 decimal digits!), making this a somewhat strict estimate, both in theory and in practice. Furthermore, the scaling by $\sqrt{n}$ is not justified if the interval is divided into $n - 1$ panels (see Equation 2.13).

2.6 Gallaher's Adaptive Quadrature Procedure

In a 1967 paper [30], Gallaher presents a recursive adaptive quadrature routine based on the midpoint rule. In this algorithm, the interval is divided symmetrically into three sub-intervals with the width h_c of the central sub-interval chosen randomly in

$$h_c \in \left[\frac{1}{6}h_k, \frac{1}{2}h_k\right], \quad h_k = (b_k - a_k).$$

The integrand $f(x)$ is evaluated at the center of each sub-interval and used to compute the midpoint rule therein (see Figure 2.5).

$$\int_a^b f(x)\, dx \approx \frac{h_k - h_c}{2}\left(f(x_1) + f(x_3)\right) + h_c f(x_2)$$

where the x_i, $i = 1\ldots3$ are the centers of each of the three panels. If the error estimate is larger than the required tolerance, the interval is trisected into the three panels used to compute the midpoint rule, thus re-using the computed function values. Since the error of the midpoint rule is known to be

$$\frac{(b_k - a_k)^3}{24} f^{(2)}(\xi), \quad \xi \in [a_k, b_k], \tag{2.15}$$

the local integration error can be estimated by computing the second divided difference of $f(x)$ over the three values of $f(x)$ in the center of the sub-intervals, f_1, f_2 and f_3:

$$\frac{(b_k - a_k)^3}{24} f^{(2)}(\xi) \approx (f_1 - 2f_2 + f_3) \frac{2(b_k - a_k)^3}{3(b_k - a_k + h_c)^2} \qquad (2.16)$$

Gallaher, however, uses the expression

$$\varepsilon = 14.6 \left| f_1 - 2f_2 + f_3 \right| \frac{b_k - a_k - h_c}{2} \sqrt{3}^d. \qquad (2.17)$$

In which the constant 14.6 is determined empirically.

Gallaher shows that this error estimate works well for a number of integrands and gives a justification for using $\sqrt{3}^d$ as opposed to 3^d as a scaling factor, stating:

> "The assumption here is that error contributed by the individual panels is random and not additive, thus the error from three panels is assumed to be $\sqrt{3}$ (not 3) times the error of one panel."

similar to the derivation in Section 2.4.

2.7   Lyness' SQUANK Integrator

In 1969, Lyness [61] publishes the first rigorous analysis of McKeeman's adaptive integrator (see Section 2.3). He suggests using the absolute local error instead of the globally relative local error, bisection instead of trisection and includes the resulting factor of 15 in the error estimate[6]:

$$\varepsilon_k = \left| S^{(1)}[a_k, b_k] - S^{(2)}[a_k, b_k] \right| \frac{2^d}{15} \qquad (2.18)$$

He further suggests using Romberg extrapolation to compute the five-node Newton-Cotes formula from the two Simpson's approximations[7]:

$$NC_5^{(1)}[a, b] = \frac{1}{15} \left(16 S^{(2)}[a, b] - S^{(1)}[a, b] \right). \qquad (2.19)$$

---

[6]Note that McKeeman's original error estimate was off by a factor of 80 (see Equation 2.7). The factor of 15 comes from using bisection instead of trisection.

[7]Interestingly enough, this was already suggested by Villars in 1956 [100] and implemented in Henriksson in 1961 [45], but apparently subsequently forgotten.

This is a departure from previous methods, in which the error estimate and the integral approximation were of the same degree, making it impracticable to relate the error estimate to the integral approximation without making additional assumptions on the smoothness of the integrand.

Note that although Lyness mentions McKeeman's [70] use of the scaling $\sqrt{3}^d$ as opposed to 3^d and notes that it *"evidently gives better results"*, he himself uses the stricter scaling 2^d. By using an *absolute* local error he avoids the problems encountered by Kuncir's error estimate (see Section 2.2, Equation 2.4), yet the depth-dependent scaling still makes the algorithm vulnerable to problems such as shown in Equation 2.8 and Equation 2.9.

Lyness later implements these improvements in the algorithm SQUANK [62], in which he also includes a guard for round-off errors: Lyness starts by showing that the un-scaled error estimate

$$S^{(1)}[a, b] - S^{(2)}[a, b] = \frac{(b-a)^5}{3072} f^{(4)}(\xi), \quad \xi \in [a, b] \tag{2.20}$$

should decrease by a factor of 16 after subdivision for constant $f^{(4)}(x)$, $x \in [a, b]$, *i.e.*

$$\frac{S^{(1)}[a, \frac{a+b}{2}] - S^{(2)}[a, \frac{a+b}{2}]}{S^{(1)}[a, b] - S^{(2)}[a, b]} = \frac{S^{(1)}[\frac{a+b}{2}, b] - S^{(2)}[\frac{a+b}{2}, b]}{S^{(1)}[a, b] - S^{(2)}[a, b]} = \frac{1}{16}.$$

If $f^{(4)}(x)$ is of constant sign for $x \in [a, b]$, then the un-scaled error measure should at least *decrease* after subdivision. Therefore, if this estimate *increases* after subdivision, Lyness assumes that one of the following has happened:

1. There is a zero of $f^{(4)}(x)$ in the interval an the error estimate is therefore not valid,

2. The error estimate is seriously affected by rounding errors in the integrand.

The first case is regarded by Lyness to be rare enough to be safely ignored. Therefore, if the error estimate increases after subdivision, significant round-off error is assumed and the tolerance is adjusted accordingly. If in latter intervals the error is smaller than the required tolerance, the tolerance is re-adjusted until the original tolerance is met. If this is not possible, then the user is informed that something may have gone horribly wrong.

It should be noted that this noise detection does not take discontinuities or singularities into account. These may have the same effect on consecutive error estimates (*i.e.* the estimate increases after subdivision), yet will be treated as noise in the integrand.

In a 1975 paper, Malcolm and Simpson [66] present a *global* version of SQUANK called SQUAGE (Simpson's Quadrature Used Adaptively Global Error) along the lines of Algorithm 2, yet without the noise detection described above. They use the same error estimate (Equation 2.18), albeit neglect to apply the Romberg extrapolation[8] and use $S^{(2)}[a_k, b_k]$ to approximate the integral. Their results show that for the same quadrature rule and local error estimate, the global, non-recursive approach (Algorithm 2) is more efficient in terms of required function evaluations than the local, recursive approach (Algorithm 1), but at the expense of larger memory requirements.

In 1977, Forsythe, Malcolm and Moler [29] publish the recursive quadrature routine QUANC8, which uses essentially the same basic error estimate as Lyness (Equation 2.18), yet using Newton-Cotes rules over 9 points, resulting in a different scaling factor:

$$\varepsilon_k = \left| NC_9^{(1)}[a_k, b_k] - NC_9^{(2)}[a_k, b_k] \right| \frac{2^d}{1023}. \tag{2.21}$$

Analogously to Equation 2.19, the two quadrature rules are combined using Romberg extrapolation to compute an approximation of degree 11:

$$\int_a^b f(x)\,dx \approx \frac{1}{1023} \left(1024 NC_9^{(2)}[a, b] - NC_9^{(1)}[a, b] \right)$$

which is used as the approximation to the integral. This routine was integrated into MATLAB as quad8, albeit without the Romberg extrapolation, and has since been replaced by quadl (see Section 2.20).

---

[8]In their paper, Malcolm and Simpson state (erroneously) that Lyness' SQUANK uses $S^{(2)}[a, b]$ as its approximation to the integral and, as their results suggest, $S^{(2)}[a, b]$ was also used in their implementation thereof. This omission, however, has no influence on their results or the conclusions they draw in their paper.

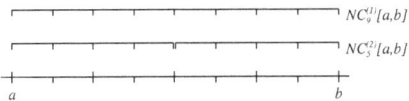

Figure 2.6: O'Hara and Smith's first error estimate computes the difference between a 9-point Newton-Cotes rule $(NC_9^{(1)}[a, b])$ and two 5-point Newton-Cotes rules $(NC_5^{(2)}[a, b])$.

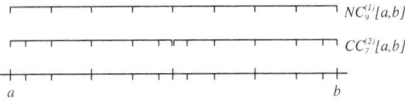

Figure 2.7: O'Hara and Smith's [77] second error estimate computes the difference between a 9-point Newton-Cotes rule $(\mathsf{NC}_9^{(1)}[a, b])$ and two 7-point Clenshaw-Curtis rules $(\mathsf{CC}_7^{(2)}[a, b])$.

2.8 O'Hara and Smith's Clenshaw-Curtis Quadrature

In 1969, O'Hara and Smith [77] publish a recursive adaptive quadrature routine based on Clenshaw-Curtis quadrature rules [10]. Their algorithm uses a cascade of error estimates. First, the error estimate

$$\varepsilon_k^{(1)} = \left| \mathsf{NC}_9^{(1)}[a_k, b_k] - \mathsf{NC}_5^{(2)}[a_k, b_k] \right| \tag{2.22}$$

is computed using 9 equidistant function evaluations in the interval (see Figure 2.6). This and subsequent error estimates are compared with a local tolerance τ_k, which will be explained later on.

If the requested tolerance is not met, the interval is sub-divided. Otherwise, a second error estimate is computed using

$$\varepsilon_k^{(2)} = \left| \mathsf{NC}_9^{(1)}[a_k, b_k] - \mathsf{CC}_7^{(2)}[a_k, b_k] \right| \tag{2.23}$$

where $\mathsf{CC}_7^{(2)}[a, b]$ is the Clenshaw-Curtis rule over 7 points applied over 2 sub-intervals of $[a, b]$. The evaluation of this rule requires the evaluation of only 4 additional function values since the points $a + \frac{1}{4}(b - a)$ and

$b - \frac{1}{4}(b - a)$ in the Clenshaw-Curtis rule overlap with already evaluated points in the Newton-Cotes rule (see Figure 2.7).

If this error estimate is larger than the required tolerance, the interval is subdivided. Otherwise, a final, third error estimate is computed using

$$\varepsilon_k^{(3)} = \frac{32}{(6^2 - 9)(6^2 - 1)} \left[\left| \sum_{i=1}^{7}{}'' (-1)^{i-1} f_{l,i} \right| + \left| \sum_{i=1}^{7}{}'' (-1)^{i-1} f_{r,i} \right| \right] \quad (2.24)$$

where Σ'' denotes a sum in which first and last terms are halved and where the $f_{l,i}$ and $f_{r,i}$ are the values of the integrand evaluated at the nodes of the 7-point Clenshaw-Curtis quadrature rules over the left and right halves of the interval respectively. These sums are, as we will see later (Equation 2.28), the approximated Chebyshev coefficients $\tilde{c}_6$ of the integrand over the left and right half of the interval.

If this third error estimate is also below the prescribed tolerance, then $CC_7^{(2)}[a_k, b_k]$ is used as an approximation to the integral. Otherwise, the interval is bisected.

The scaled tolerance τ_k is computed and updated such that it represents one tenth of the global tolerance τ minus the accumulated error estimates of the previously converged intervals. That is, as the algorithm recurs towards the leftmost interval, τ_1 is simply 0.1τ. Once the leftmost interval has converged, with an estimated error ε_1, the τ_2 of the next interval is set to $0.1(\tau - \varepsilon_1)$. For the kth interval, this can be written recursively as

$$\tau_k = 0.1 \left(\tau - \sum_{i=1}^{k-1} \varepsilon_i \right). \quad (2.25)$$

Assuming that $\varepsilon_k \leq \tau_k$, the local tolerance in the kth interval is bounded by

$$\left(\frac{9}{10} \right)^{k-1} \frac{\tau}{10} \leq \tau_k \leq 0.1\tau.$$

Using this scheme, the global error estimate after processing k intervals should be bounded by

$$\varepsilon \leq \sum_{i=1}^{k} \tau_i = \left(1 - \left(\frac{9}{10} \right)^k \right) \tau.$$

note that as $k \to \infty$, the error estimate becomes more and more restrictive. This can become a problem if we have a singularity at the right end of the interval and can lead the algorithm to fail. In general, this type of local tolerance is problematic since it requires the algorithm to allot a certain part of the error tolerance before knowing how the rest of the function will or may behave. This allocation can be either overly generous (*i.e.* for a singularity/discontinuity on the right) or overly restrictive (*i.e.* for a singularity/discontinuity on the left).

The first two error estimates (Equation 2.22 and Equation 2.23) are similar to those used by Kuncir, McKeeman and Lyness, yet differ in that the difference between estimates of *different degree* are computed, *i.e.* they use the difference between a rule of degree 9 and one of degree 5 or one of degree 7 respectively. It is therefore no longer possible to formulate the error estimate as a multiple of the actual error for "sufficiently smooth" integrands without making additional assumptions on the integrand itself.

The third error estimate (Equation 2.24) is derived by O'Hara and Smith in [76] based on the error estimation used by Clenshaw and Curtis in [10]. Consider, as is done in [10], that the function $f(x)$ in $x \in [a, b]$ has a Chebyshev expansion

$$f\left(\frac{a+b}{2} + \frac{b-a}{2}x\right) = \sum_{i=0}^{\infty}{}' c_i T_i(x), \quad x \in [-1, 1] \qquad (2.26)$$

in the interval $[a, b]$, where Σ' denotes a sum in which first term is halved and the $T_i(x)$ are the Chebyshev polynomials satisfying

$$T_0(x) = 1, \quad T_1(x) = x, \quad T_k(x) = 2xT_{k-1}(x) - T_{k-2}(x).$$

The integral of $f(x)$ can therefore be written as

$$\begin{aligned}
\int_a^b f(x)\,\mathrm{d}x &= (b-a)\sum_{i=0}^{\infty}{}' c_i \int_{-1}^1 T_i(x)\,\mathrm{d}x \\
&= (b-a)\left[c_0 - \frac{2}{3}c_2 - \frac{2}{15}c_4 - \cdots \right. \\
&\qquad \left. - \frac{2}{(2i-1)(2i+1)}c_{2i} - \cdots \right].
\end{aligned} \qquad (2.27)$$

The coefficients c_i are unknown, yet they can be approximated using

$$\tilde{c}_i = \frac{2}{n} \sum_{j=0}^{n}{}'' f(x_j) \cos \frac{\pi i j}{n} \tag{2.28}$$

where the x_j, $j = 0 \ldots n$ are the $n+1$ Chebyshev nodes in $[a, b]$:

$$x_j = a + \frac{(b-a)}{2} \left(\cos \frac{\pi j}{n} + 1 \right).$$

Inserting the expansion in Equation 2.26 into Equation 2.28 and using the orthogonality of the Chebyshev polynomials, the approximate coefficients can be expressed, for even n, as

$$\tilde{c}_i = c_i + c_{2n-i} + c_{2n+i} + c_{4n-i} + c_{4n+i} + \ldots \tag{2.29}$$

Using the $n+1$ approximated coefficients $\tilde{c}_i$, and integrating over the Chebyshev polynomials as in Equation 2.27, we can approximate the integral as we did in Equation 2.27, which is what the $n+1$ point Clenshaw-Curtis rule computes:

$$\mathsf{CC}_{n+1}^{(1)}[a, b] =$$

$$(b-a) \left[\tilde{c}_0 - \frac{2}{3}\tilde{c}_2 - \frac{2}{15}\tilde{c}_4 - \cdots - \frac{2}{(n-1)(n+1)}\tilde{c}_n \right] \tag{2.30}$$

for even n. Now, inserting Equation 2.29 into Equation 2.30 and subtracting it from Equation 2.27, assuming that the coefficients c_i may be neglected for $i \geq 3n$, we can compute

$$\int_a^b f(x) \, \mathrm{d}x - \mathsf{CC}_{n+1}^{(1)}[a, b] =$$

$$(b-a) \left[-2c_{2n} + \frac{2}{3}(c_{2n-2} + c_{2n+2}) + \ldots \right.$$

$$+ \frac{2}{(n-3)(n-1)}(c_{n+2} + c_{3n-2})$$

$$\left. - \sum_{i=1}^{n-1} \frac{2c_{n+2i}}{(n+2i-1)(n+2i+1)} \right] \tag{2.31}$$

as is done in [10]. O'Hara and Smith, however, collect the terms in Equation 2.31 and write

$$\int_a^b f(x)\,dx - \mathsf{CC}_{n+1}^{(1)}[a,b] =$$

$$(b-a)\left[\frac{16n}{(n^2-1)(n^2-9)}c_{n+2} + \frac{32n}{(n^2-9)(n^2-25)}c_{n+4} + \cdots \right.$$

$$+ \; \frac{16(n/2-1)n}{3(2n-1)(2n-3)}c_{2n-2} - \left(2 + \frac{2}{4n^2-1}\right)c_{2n}$$

$$+ \; \left. \left(\frac{2}{3} - \frac{2}{(2n+1)(2n+3)}\right)c_{2n+2} + \cdots \right]. \qquad (2.32)$$

They note that for most regular functions, the magnitude of the coefficients c_i of its Chebyshev series fall to zero exponentially as $i \to \infty$ and therefore, the first term in Equation 2.32 is often larger than the sum of the following terms.

They find that if they define the higher-order $|c_{2i}|$, $i > n+1$ in terms of $|c_{n+2}|$ using the recurrence relation $|c_{i+2}| = K_n|c_i|$, then they can define K_n for different n such that the first term of Equation 2.32 dominates the series. For the 7-point Clenshaw-Curtis rule, this value is $K_6 = 0.12$. If the relation $|c_{i+2}| \le K_n|c_i|$ holds, then the error is bounded by twice the first term of Equation 2.32

$$\left| \int_a^b f(x)\,dx - \mathsf{CC}_{n+1}^{(1)}[a,b] \right| \le (b-a)\frac{32n}{(n^2-1)(n^2-9)}|c_{n+2}|.$$

However, we do not know c_{n+2}, yet since we assume that the magnitude of the coefficients decays, we can assume that

$$|c_{n+2}| < |c_n| \approx \frac{1}{2}|\tilde{c}_n|$$

and use $\frac{1}{2}|\tilde{c}_n|$. Since $|c_n|$ might be "accidentally small", they suggest, in [76], as an error estimate

$$\varepsilon = (b-a)\frac{16n}{(n^2-1)(n^2-9)} \max\left\{|\tilde{c}_n|, 2K_n|\tilde{c}_{n-2}|, 2K_n^2|\tilde{c}_{n-4}|\right\}. \qquad (2.33)$$

In their algorithm, however, since after passing the first two error estimates the function is assumed to be sufficiently smooth, the third error estimate Equation 2.24 uses only $|\tilde{c}_n|$ instead of $\max\{\cdots\}$.

2.9   De Boor's CADRE Error Estimator

In 1971, de Boor [14] publishes the integration subroutine CADRE. The algorithm, which follows the scheme in Algorithm 1, generates a Romberg T-table [3] with

$$T_{\ell,i} = T_{\ell,i-1} + \frac{T_{\ell,i-1} - T_{\ell-1,i-1}}{4^i - 1} \qquad (2.34)$$

in every interval. The entries of the T-table are equivalent to quadratures of the type

$$T_{\ell,i} = \frac{1}{b - a} Q_{2i+1}^{(2^{\ell-i})}[a, b]. \qquad (2.35)$$

The entries in the T-table are used to decide whether to extend the table or bisect the interval[9]. After adding each ℓth row to the table, the decision is made as follows:

1. If $\ell = 1$ and $T_{1,0} = T_{0,0}$, assume that the integrand is a straight line. This is verified by evaluating the integrand at four random[10] points in the interval. If the computed values differ from the expected value by more than $\varepsilon_{\mathsf{mach}}|\hat{I}_k|$, where $|\hat{I}_k|$ is the approximation $T_{\ell,1}$ computed over the absolute function values, then the interval is bisected. Otherwise $T_{\ell,0}$ is returned as an approximation of the integral.

2. If $\ell > 1$, the ratios
$$R_i = \frac{T_{\ell-1,i} - T_{\ell-2,i}}{T_{\ell,i} - T_{\ell-1,i}} \qquad (2.36)$$
for $i = 0 \ldots \ell - 2$ are computed.

3. If $R_0 = 4 \pm 0.15$, the function is assumed to be sufficiently smooth and "cautious extrapolation" is attempted: $(b_k - a_k)T_{\ell,i}$ is used as an approximation to the integral for the smallest $i \leq \ell$ for which

$$\varepsilon_k = \left| (b_k - a_k) \frac{T_{\ell,i-1} - T_{\ell-1,i-1}}{4^i - 1} \right| \qquad (2.37)$$

---

[9]Thus making it the first doubly-adaptive quadrature algorithm known to the author, although this title is usually bestowed upon other authors, e.g. Cools and Haegmans by [21] or Oliver by [24].

[10]The "random" points are hard-coded as [0.71420053, 0.34662815, 0.843751, 0.12633046] for the interval [0, 1].

is smaller than the local tolerance τ_k.

4. If $R_0 = 2 \pm 0.01$, the integrand is assumed to have a jump discontinuity and the error is bounded by the difference between the two previous estimates, $T_{\ell,0}$ and $T_{\ell-1,0}$. If $\tau_k < |T_{\ell,0} - T_{\ell-1,0}|$ the interval is bisected. Otherwise, $T_{\ell,0}$ is returned as an approximation to the integral.

5. If R_0 is within 10% of the R_0 computed at the previous level $\ell - 1$ and $R_0 \in (1,4)$, assume that the interval has an integrable singularity of the type

$$f(x) = (x - \xi)^\alpha g(x) \qquad (2.38)$$

where ξ is near the edges of $[a_k, b_k]$ and $\alpha \in (-1, 1)$. If this is the case, R_0 should be $\approx 2^{\alpha+1}$. The exponent α (or rather, the value $2^{\alpha+1}$) is estimated from successive R_i and used to recompute the T-table where *"cautious extrapolation"* is attempted as above, yet interleaving the recursion

$$T_{\ell,i} = T_{\ell,i-1} + \frac{T_{\ell,i-1} - T_{\ell-1,i-1}}{2^{\alpha+i} - 1} \qquad (2.39)$$

where necessary. As with above, the first entry to satisfy the error requirement is returned as an approximation to the integral. If no such estimate is found, the interval is bisected.

6. If the previous tests have failed, the *"noise level"* of the integrand in the interval is approximated by evaluating the function at four *"random"* nodes inside the interval and comparing the functions values to the line through the endpoints of the interval. If each of these differences is smaller than the required tolerance the last estimate of the integrand considered ($T_{\ell,i}$) is returned.

7. Otherwise, if $\ell < 5$ or the rate of decrease of the error computed from the last two T-table rows indicates that the scheme will converge for $\ell < 10$, another row is added to the table.

8. Otherwise, the interval is bisected.

The local tolerance τ_k for each kth interval is computed as

$$\tau_k = \max\left\{ (b_k - a_k)\varepsilon_{\mathsf{mach}}|T_{\ell,1}|, 2^{-d} \max\left\{ \tau_{\mathsf{abs}}, \tau_{\mathsf{rel}} \left| \hat{I} \right| \right\} \right\} \qquad (2.40)$$

where τ_{abs} and τ_{rel} are the absolute and relative error requirements given by the user and $\hat{I}$ is the current approximation to the *global* integral. This local tolerance, which is scaled with 2^{-d} where d is the recursion depth, is in essence the strict tolerance first used by Kuncir (see Section 2.2), yet with a guard for error estimates below machine precision relative to the *local* integral itself, similar to that of McKeeman (see Section 2.4).

The rationale for using the ratios R_i (Equation 2.36) is based on the observation that, given the correspondence in Equation 2.35, the error of each entry of the T-table is, for sufficiently smooth integrands,

$$\frac{1}{b-a} \int_a^b f(x)\,dx - T_{\ell,i} \approx \kappa_i \left(2^{-(\ell-i)}\right)^{2i+2}. \tag{2.41}$$

The ratio R_i can therefore be re-written as

$$
\begin{aligned}
R_i &= \frac{\kappa_i \left(2^{-(\ell-i-1)}\right)^{2i+2} - \kappa_i \left(2^{-(\ell-2-i)}\right)^{2i+2}}{\kappa_i \left(2^{-(\ell-i)}\right)^{2i+2} - \kappa_i \left(2^{-(\ell-1-i)}\right)^{2i+2}} \\
&= \frac{2^{2i+2} - 4^{2i+2}}{1 - 2^{2i+2}} \\
&= 4^{i+1}. \tag{2.42}
\end{aligned}
$$

If this condition is actually satisfied (approximately), then de Boor considers it safe to assume that the difference between the two approximations $T_{\ell,i-1}$ and $T_{\ell,i}$ is a good bound for the error of $T_{\ell,i}$, as is computed in Equation 2.37[11].

Assuming that the integrand does have the form described in Equation 2.38, with $\alpha \in (-1, 1)$ then, as Lyness and Ninham [65] have shown, assuming $f(x)$ is $2k + 1$ times continuously differentiable,

$$\frac{1}{b-a} \int_a^b f(x)\,dx - T_{\ell,0} =$$

$$\sum_{1 \leq i \leq 2k+1-\alpha} A_i h^{i+\alpha} + \sum_{i=1}^{k} B_i h^{2i} + \mathcal{O}(h^{2k+1}), \quad h = 2^{-\ell}. \tag{2.43}$$

---

[11]Remember that $T_{\ell,i} = T_{\ell,i-1} + (T_{\ell,i-1} - T_{\ell-1,i-1})/(4^i - 1)$.

Thus, when constructing the T-table, we need to eliminate not only the terms in h^{2i}, as is usually done, but also the terms in $h^{i+\alpha}$, using

$$T_{\ell,i} = T_{\ell,i-1} + \frac{T_{\ell,i-1} - T_{\ell-1,i-1}}{2^{i+\alpha} - 1} \qquad (2.44)$$

interleaved with the "normal" recurrence relation (Equation 2.34). As with the "normal" T-table, the error estimate is computed as in Equation 2.37 as the difference between the last two elements in this "special" T-table.

Another interesting feature of this algorithm is that it returns, along with the integral and error estimates, a value IFLAG which can have any of the following values (and their descriptions from [14]):

1: *"All is well,"*

2: *"One or more singularities were successfully handled,"*

3: *"In one or more sub-intervals the local estimate was accepted merely because the error estimate was small, even though no regular behavior could be recognized,"*

4: *"Failure, overflow of stack,"*

5: *"Failure, too small a sub-interval is required, which may be due to too much noise in the function relative to the given error requirement or due to a plain ornery integrand."*

Regarding these values, de Boor himself states:

> *"A very cautious man would accept* CADRE *[the returned value of the integral] only if* IFLAG *is 1 or 2. The merely reasonable man would keep the faith even if* IFLAG *is 3. The adventurous man is quite often right in accepting* CADRE *even if* IFLAG *is 4 or 5."*

2.10   Rowland and Varol's Modified Exit Procedure

In 1972, Rowland and Varol [88] publish an error estimator based on Simpson's compound rule. In their paper, they show that the *"stopping inequality"*

$$\left| S^{(m)}[a, b] - S^{(2m)}[a, b] \right| \geq \left| S^{(2m)}[a, b] - \int_a^b f(x) \, dx \right|$$

as it is used by Lyness with $m = 1$ (see Section 2.7) is valid if $f^{(4)}(x)$ is of constant sign for $x \in [a, b]$. They also show that under certain conditions there exists an integer m_0 such that the inequality is valid for all $m \geq m_0$.

Similarly to de Boor (see Section 2.9), they note that for the compound Simpson's rule

$$\frac{S^{(m)}[a, b] - S^{(2m)}[a, b]}{S^{(2m)}[a, b] - S^{(4m)}[a, b]} \approx 2^{2q} \qquad (2.45)$$

holds, where usually $q = 2$. This condition is used to test if m is indeed large enough, much in the same way as de Boor's CADRE does (Equation 2.36 in Section 2.9) to test for regularity. If this condition is more or less satisfied[12] for any given m, then they suggest using

$$\varepsilon_k = \frac{\left(S^{(2m)}[a_k, b_k] - S^{(4m)}[a_k, b_k] \right)^2}{\left| S^{(m)}[a_k, b_k] - S^{(2m)}[a_k, b_k] \right|}. \qquad (2.46)$$

This error estimate can be interpreted as follows: Let us assume that

$$e_m = \left| S^{(m)}[a, b] - S^{(2m)}[a, b] \right| \qquad (2.47)$$

is, for a sufficiently smooth integrand or sufficiently large m, an estimate of the error of $S^{(m)}[a, b]$. This error estimate is in itself somewhat impracticable, since to compute it, one has to evaluate the more precise compound rule $S^{(2m)}[a, b]$, and if we are to compute this rule, we would rather use it instead of $S^{(m)}[a, b]$ as our estimate to the integral. If we assume, however, that the error estimates decrease at a constant rate[13] r such that

$$\begin{aligned} e_{2m} &= r e_m \\ r &= \frac{e_{2m}}{e_m}, \end{aligned} \qquad (2.48)$$

---

[12]Since their paper does not include an implementation, no specification is given to how close to a power of two this ratio has to be.

[13]If the error of the compound rule behaves as in Equation 1.2, then this assumption is indeed valid.

then we can *extrapolate* the error of $S^{(4m)}[a, b]$ using

$$
\begin{aligned}
e_{4m} &= re_{2m} \\
&= \frac{e_{2m}^2}{e_m}
\end{aligned}
\tag{2.49}
$$

which is exactly what is computed in Equation 2.46.

This error estimate can also be explained in terms of the Romberg T-table, where $S^{(m)}[a, b]$, $S^{(2m)}[a, b]$ and $S^{(4m)}[a, b]$ are the last three entries in the second column, $T_{\ell-2,1}$, $T_{\ell-1,1}$ and $T_{\ell,1}$. If we assume that the relation in Equation 2.45 holds with $q = 2$, then the error estimate is equivalent to

$$
\varepsilon_k = \frac{1}{16} \left| T_{\ell,1} - T_{\ell-1,1} \right|,
$$

which, except for a constant factor of $\frac{16}{15}$, is the same error estimate used by de Boor for regular integrands (see Section 2.9, Equation 2.37).

The main differences between both algorithms are that

1. Rowland and Varol only consider the second column of the Romberg T-table,

2. CADRE tests for singularities and discontinuities explicitly, whereas Rowland and Varol's error estimate tries to catch them implicitly by extrapolating the convergence rate directly (see Equation 2.45).

A similar approach is taken by Venter and Laurie [99], where instead of using compound Simpson's rules of increasing multiplicity, they used a sequence of stratified quadrature rules, described by Laurie in [58]. In their algorithm, the sequence of quadratures

$$
Q_1[a, b], Q_3[a, b], Q_7[a, b], \ldots, Q_{2^i-1}[a, b]
$$

is computed and the differences

$$
E_i = \left| Q_{2^i-1}[a, b] - Q_{2^{i+1}-1}[a, b] \right|
$$

are used to extrapolate the error of the highest-order (*i*th) quadrature rule:

$$
\varepsilon_k = \frac{E_{i-1}^2}{E_{i-2}}.
\tag{2.50}
$$

The resulting algorithm, which is implemented as an adaptation of QUAD-PACK's `QAG` (see Section 2.17), follows the global scheme in Algorithm 2. The interval is bisected if $i = 8$ or the ratio

$$\text{hint} = \frac{E_{i-1}}{E_{i-2}} \tag{2.51}$$

is larger than 0.1, an empirically determined threshold.

In [98], Venter notes that the error estimates usually converge *faster* than linearly. If we assume that the error of the ith quadrature rule behaves as

$$Q_{2^i-1}[a,b] - \int_a^b f(x)\,\mathrm{d}x \approx \kappa_{2^i}(b-a)^{2^i+1}$$

and that the coefficients κ_{2^i} are of more or less the same magnitude, then the successive error estimates can be expected to decrease as

$$\frac{E_i}{E_{i-1}} \sim 2^{-(i+1)}.$$

A similar approach, yet using nested Clenshaw-Curtis rules, had already been suggested in 1992 by Pérez-Jordá *et al.* [81]. They use the same extrapolation as Venter and Laurie Equation 2.50, yet the errors E_{i-1} and E_{i-2} are approximated using

$$E_{i-1} = |Q_{2^i-1}[a,b] - Q_{2^{i-1}-1}[a,b]| \tag{2.52}$$
$$E_{i-2} = |Q_{2^i-1}[a,b] - Q_{2^{i-2}-1}[a,b]| \tag{2.53}$$

resulting in the error estimate

$$\varepsilon = \frac{(Q_{2^i-1}[a,b] - Q_{2^{i-1}-1}[a,b])^2}{|Q_{2^i-1}[a,b] - Q_{2^{i-2}-1}[a,b]|}.$$

Since their algorithm is non-adaptive, though, it is only mentioned in passing.

2.11   Oliver's Doubly-Adaptive Quadrature Routine

In 1972, Oliver [79] presents a recursive doubly-adaptive Clenshaw-Curtis quadrature routine using an extension of the error estimate of O'Hara and

Smith (see Section 2.8). The quadrature routine is doubly adaptive in the sense that for each interval after the application of an $(n + 1)$-point Clenshaw-Curtis rule, the algorithm make a decision whether to double the order of the quadrature rule to $2n + 1$ points or to subdivide the interval.

Instead of assuming a constant K_n such that

$$|c_{i+2}| \leq K_n |c_i|$$

where the c_i are the Chebyshev coefficients (see Equation 2.26) of the integrand, as do O'Hara and Smith, such that the first term in the error expansion (Equation 2.32) dominates the sum of all following terms, Oliver approximates the smallest rate of decrease of the coefficients as

$$K = \max \left\{ \left| \frac{\tilde{c}_n}{\tilde{c}_{n-2}} \right|, \left| \frac{\tilde{c}_{n-2}}{\tilde{c}_{n-4}} \right|, \left| \frac{\tilde{c}_{n-4}}{\tilde{c}_{n-6}} \right| \right\} \qquad (2.54)$$

where the $\tilde{c}_i$ are the computed Chebyshev coefficients (see Equation 2.28). He also pre-computes a number of convergence rates $K_n(\sigma)$, which are the rates of decay required such that, for n coefficients, σ times the first term of the error expansion in Equation 2.32 dominates the sum of the remaining terms. If K is less than any $K_n(\sigma)$ for $\sigma = 2$, 4, 8 or 16, then the error estimate

$$\varepsilon_k = \sigma \frac{16n}{(n^2 - 1)(n^2 - 9)} \max \left\{ K|\tilde{c}_n|, K^2|\tilde{c}_{n-2}|, K^3|\tilde{c}_{n-4}| \right\}, \qquad (2.55)$$

which is consistent with Equation 2.33 by O'Hara and Smith, is used.

If $K \leq K_n(16)$, then ε_k Equation 2.55 is used as an error estimate. Otherwise, since the previous extrapolations might have been unreliable, the difference between the two previous estimates

$$\left| CC_{n+1}^{(1)}[a, b] - CC_{n/2+1}^{(1)}[a, b] \right|$$

is used. If this estimate does not exist (*i.e.* we have just started on this interval), then the interval is immediately subdivided.

The error estimate is compared to the required local tolerance

$$\tau_k = \left(\tau - \sum_{i=1}^{k-1} \varepsilon_i \right) \max \left\{ \frac{2(b_k - a_k)}{b - a_k}, 0.1 \right\} \qquad (2.56)$$

which is similar to the local tolerance of O'Hara and Smith (Equation 2.25), yet with a relaxation when the interval is larger than one tenth of the remaining integration interval.

If ε_k exceeds the required local tolerance τ_k, the computed rate of decrease K is compared to a pre-computed limit K_n^*. This limit is defined by Oliver in [78] as the rate of decrease of the Chebyshev coefficients as of which it is preferable to subdivide the interval as opposed to doubling the order of the quadrature rule. Therefore, if $K > K_n^*$, the interval is subdivided, otherwise the order of the Clenshaw-Curtis quadrature rule is doubled.

2.12   Gauss-Kronrod Type Error Estimators

In 1973 both Patterson [80] and Piessens [82] publish adaptive quadrature routines based on Gauss quadrature rules and their Kronrod extensions [54].

Piessens' algorithm computes both a n-point Gauss quadrature $\mathsf{G}_n[a, b]$ of degree $2n - 1$ and its $2n + 1$-point Kronrod extension $\mathsf{K}_{2n+1}[a, b]$ of degree $3n + 1$. The Kronrod extension is used as an approximation to the integral and an the error estimate

$$\varepsilon_k = |\mathsf{G}_n[a_k, b_k] - \mathsf{K}_{2n+1}[a_k, b_k]| \tag{2.57}$$

is computed. Notice the absence of a scaling factor relative to the recursion depth: it is not necessary since this is the first algorithm published adhering to the heap-based scheme shown in Algorithm 2. The global error is computed as the sum of the local errors ε_k and subdivision of the interval with the largest local error estimate continues until the sum of the local error estimates (hence the global error estimate) is below the required tolerance. This avoids any problems related to having to estimate a local tolerance, such as those observed in Equation 2.8 and Equation 2.9.

Patterson's integrator takes a different approach, starting with a 3-point Gauss quadrature rule and successively extending it using the Kronrod scheme [54] to 7, 15, 31, 63, 127 and 255 nodes, resulting in quadrature rules of degree 5, 11, 23, 47, 95, 191 and 383 respectively, until the difference between two successive estimates is below the required local tolerance:

$$\varepsilon_k = |\mathsf{K}_n[a_k, b_k] - \mathsf{K}_{2n+1}[a_k, b_k]| / |\hat{I}| \qquad (2.58)$$

where $\mathsf{K}_n[a, b]$ is the Kronrod extension over n nodes and $\mathsf{K}_{2n+1}[a, b]$ its extension over $2n + 1$ nodes. $\hat{I}$ is an initial approximation of the global integral generated by applying successive Kronrod extensions to the whole interval before subdividing.

The interval is only subdivided if the error estimate for the highest-degree rule is larger than the requested tolerance. Since it is not possible to re-use points after subdivision of the interval, subdivision itself is regarded as a "rescue operation" to be used only as a last resort.

Notice here also the lack of a scaling factor relative to the recursion depth, although, as opposed to Piessens' algorithm, this algorithm works recursively (as in Algorithm 1). If many subdivisions occur, then the sum of the local error estimates, each smaller the global tolerance, may not necessarily be itself smaller than the global tolerance. The sum of the local error estimates is therefore also returned such that the user may himself/herself verify the quality of the integration.

In both algorithms the error is approximated using quadrature rules of different degree, ss with the estimator of O'Hara and Smith (see Section 2.8), thus rendering an analytical interpretation somewhat difficult.

2.13   The SNIFF Error Estimator

In 1978, Garribba et al. [35] publish SNIFF, a Self-tuning Numerical Integrator For analytical Functions. Their algorithm is one of the few exceptions which differs, in its general structure, from both Algorithm 1 and Algorithm 2: Since it is based on Gauss-Legendre quadrature rules, and therefore no function evaluations can be recycled upon interval subdivision, the authors decided on a scheme along the lines of Algorithm 3, similar to the algorithm of Morrin (see Section 2.1).

The interval subdivision works by first considering the entire interval $[a, b]$ and computing the integral and the error estimate. If the requested tolerance is not satisfied, the interval is reduced by

$$[a_{k+1}, b_{k+1}] = [a_k, a_k + h_k], \quad h_k = \frac{b_k - a_k}{m_k}$$

where m_k is chosen dynamically, as we will see later. Once an interval $[a_k, b_k]$ has converged, the next interval is chosen as

$$[a_{k+1}, b_{k+1}] = [b_k, \min\{b_k + h_k, b\}], \quad h_k = \frac{b_k - a_k}{m_k}$$

where, since the kth interval was accepted, $m_k \leq 1$.

For the error estimate, they start by noting that for any integration rule Q_n of degree n, the error can be written as

$$Q_n^{(1)}[a, b] - \int_a^b f(x)\,\mathrm{d}x = K_n(b - a)^{n+2} f^{(n+1)}(\xi), \quad \xi \in [a, b] \quad (2.59)$$

where the constant K_n depends on the type an number of nodes of the quadrature rule $Q_n^{(1)}[a, b]$.

If the interval is subdivided into m sub-intervals of equal width, this estimate becomes

$$Q_n^{(m)}[a, b] - \int_a^b f(x)\,\mathrm{d}x \approx K_n \frac{(b - a)^{n+2}}{m^{n+2}} \bar{f}^{(n+1)}$$

where $\bar{f}^{(n+1)}$ is the average of the m $n + 1$st derivatives $f^{(n+1)}(\xi_i)$, $\xi_k \in [a + (i - 1)h, a + ih]$, $h = (b - a)/m$ from the errors of each of the m panels of $Q_n^{(m)}[a, b]$. They combine these two estimates for $m = 1$ and $m = 2$, as is done in Romberg extrapolation, to approximate the value

$$\kappa_n := \frac{2^{n+2}\left(Q_n^{(1)}[a, b] - Q_n^{(2)}[a, b]\right)}{2^{n+2} - 1} \approx K_n(b - a)^{n+2} \bar{f}^{(n+1)}. \quad (2.60)$$

The resulting value κ_n can then be used to represent the error of any compound rule of multiplicity m:

$$Q_n^{(m)}[a, b] - \int_a^b f(x)\,\mathrm{d}x \approx \kappa_n m^{-(n+2)}. \quad (2.61)$$

Given the representation of the locally relative error of the kth interval

$$\left| Q_n^{(m)}[a_k, b_k] - \int_{a_k}^{b_k} f(x)\,\mathrm{d}x \right| \leq \tau_k \left| \int_{a_k}^{b_k} f(x)\,\mathrm{d}x \right| \qquad (2.62)$$

where τ_k is the required tolerance for that interval, they insert Equation 2.61 into the left hand side of Equation 2.62 and $\int_a^b f(x)\,\mathrm{d}x \approx Q_n^{(1)} - \kappa_n$ into its right hand side and solve for m, resulting in

$$\frac{2^{n+2} \left| Q_n^{(2)}[a_k, b_k] - Q_n^{(1)}[a_k, b_k] \right|}{\tau_k \left| 2^{n+2} Q_n^{(2)}[a_k, b_k] - Q_n^{(1)}[a_k, b_k] \right|} \leq m^{n+2} \qquad (2.63)$$

Using this relation, they compute the number of intervals m_k in which they should subdivide the domain such that the relative tolerance τ_k should be satisfied:

$$m_k = \left(\frac{2^{n+2} \left| Q_n^{(2)}[a_k, b_k] - Q_n^{(1)}[a_k, b_k] \right|}{\tau_k \left| 2^{n+2} Q_n^{(2)}[a_k, b_k] - Q_n^{(1)}[a_k, b_k] \right|} \right)^{1/(n+2)} .$$

If $m_k \leq 1$, then the interval is sufficiently converged and no further subdivision is needed. We can re-formulate this condition as an error estimate for the quadrature rule $Q_n^{(2)}$ from Equation 2.61:

$$\varepsilon_k = \frac{\left| Q_n^{(2)}[a_k, b_k] - Q_n^{(1)}[a_k, b_k] \right|}{2^{n+2} - 1} \qquad (2.64)$$

according to which, if it is larger than τ_k, the interval will be subdivided into m_k sub-intervals. This is similar to the error estimate of de Boor for the entries of the Romberg T-table (see Section 2.9, Equation 2.37).

The local tolerance τ_k for the kth interval is computed from the global relative tolerance τ as

$$\tau_k = \tau \max\left\{ 1, \left| \frac{\hat{I}_k}{Q_n^{(2)}[a_k, b_k]} \right| \right\} \left| \frac{(b_k - a_k)}{(b - a)} \right| \qquad (2.65)$$

where $\hat{I}_k$ is the running estimate of the global integral at the time that the kth interval is inspected. The factor $\hat{I}_k / Q_n^{(2)}[a_k, b_k]$ basically converts the relative tolerance into an absolute tolerance and the factor $(b_k - a_k)/(b - a)$ scales the local tolerance relative to the interval width. If we assume that the cumulative estimates $\hat{I}_k$ are correct, then the sum of the tolerances is equal to $\tau \left| \int_a^b f(x)\, dx \right|$. This is, in essence, the same as using a *globally relative* local tolerance and scaling the local error with m after each subdivision into m sub-intervals, as is done by McKeeman (see Section 2.4) and therefore shares the same problems as shown in Equation 2.8.

2.14   De Doncker's Adaptive Extrapolation Algorithm

The probably most well-known quadrature algorithm using non-linear extrapolation is published by de Doncker in 1978 [15]. The main idea of the algorithm is similar to that of the Romberg scheme: Given a basic quadrature rule $Q_n[a, b]$, the sequence

$$Q_n^{(1)}[a, b], Q_n^{(2)}[a, b], Q_n^{(4)}[a, b], \ldots, Q_n^{(2^i)}[a, b], \ldots \qquad (2.66)$$

converges exponentially[14], for large enough i, towards $\int_a^b f(x)\, dx$.

In Romberg's scheme, $Q_n^{(m)}[a, b] = T^{(m)}[a, b]$, the trapezoidal rule, and the limit of the sequence is extrapolated linearly using the the Romberg T-table. De Doncker's algorithm, however, differs in two main points:

1. The 21-point Gauss-Kronrod rule is used as the basic rule $Q_n^{(m)}[a, b]$ instead of the trapezoidal rule,

2. The non-linear ϵ-Algorithm [102] is used to extrapolate the limit of the sequence instead of the Romberg T-table.

The ϵ-Algorithm computes a sequence of approximations using the recursion

$$\epsilon_k^{(j)} = \epsilon_{k-2}^{(j+1)} + \left(\epsilon_{k-1}^{(j+1)} - \epsilon_{k-1}^{(j)} \right)^{-1}$$

---

[14]If we assume that the error of each quadrature behaves as in Equation 1.2, then the error decreases by a factor of $2^{-(n+2)}$ every time m is doubled.

where $\epsilon_0^{(j)} = Q_n^{(2^j)}[a, b]$, $j = 0 \ldots i$ are the initial approximations to the integral and $\epsilon_j^{(-1)} = 0$. This recursion is usually represented and stored in an epsilon-table

$$
\begin{array}{ccccccc}
 & \epsilon_0^{(0)} & & & & & \\
\epsilon_{-1}^{(1)} & & \epsilon_1^{(0)} & & & & \\
 & \epsilon_0^{(1)} & & \epsilon_2^{(0)} & & & \\
\epsilon_{-1}^{(2)} & & \epsilon_1^{(1)} & & \epsilon_3^{(0)} & & \\
 & \epsilon_0^{(2)} & & \epsilon_2^{(1)} & & \ddots & \\
\epsilon_{-1}^{(3)} & & \epsilon_1^{(2)} & & \ddots & & \\
 & \epsilon_0^{(3)} & & \ddots & & & \\
\epsilon_{-1}^{(4)} & \vdots & \ddots & & & &
\end{array}
$$

in which the even columns $\epsilon_{2i}^{(j)}$ represent the faster converging sequences. The last entry in the rightmost even column is then used as an estimate for the sequence limit, in this case I_i.

The use of the Gauss-Kronrod rule for adaptive quadrature is not new, as it was first introduced by Piessens in 1973 (see Section 2.12) and the use of the ϵ-Algorithm instead of the Romberg T-table was already tested by Kahaner in 1972 [50]. The combination of both, however, with the following modification, is novel.

The algorithm, as described thus far, is not yet adaptive. The main (and new) innovation is that instead of using $Q_n^{(m)}[a, b]$, de Doncker uses *approximations* $\tilde{Q}_n^{(m)}[a, b]$. Each approximation $\tilde{Q}_n^{(m)}[a, b]$ is computed by iteratively picking out the sub-interval of width greater than $h = (b - a)/m$ with the largest local error estimate

$$\varepsilon_k = |G_{10}[a_k, b_k] - K_{21}[a_k, b_k]| \tag{2.67}$$

which is the same local error estimate as first used by Piessens (see Section 2.12), and subdividing it until:

1. The sum of the local error estimates ε_k of all intervals of width larger than h is smaller than the required tolerance or

2. There are no more intervals of width larger than h left to subdivide.

In the first case, the error induced by *not* subdividing the intervals larger than h is below the required global tolerance τ and thus so should

$$\left| \tilde{Q}_n^{(m)}[a,b] - Q_n^{(m)}[a,b] \right| < \tau.$$

In the second case, all intervals are of the width h and therefore $\tilde{Q}_n^{(m)}[a,b] = Q_n^{(m)}[a,b]$.

In her original paper, de Doncker does not give any details on how the ϵ-Algorithm is applied or how the global error is estimated. The algorithm, however, was later packaged as QAGS in QUADPACK. In that implementation, the local error estimate (Equation 2.67) is replaced by the local error estimator used in the other QUADPACK-routines (see Section 2.17, Equation 2.97). Furthermore, to avoid problems in the extrapolation due to bad initial estimates, the error

$$e_k^{(j)} = \left| \epsilon_{k-2}^{(j)} - \epsilon_{k-2}^{(j+1)} \right| + \left| \epsilon_{k-2}^{(j+1)} - \epsilon_{k-2}^{(j+2)} \right| + \left| \epsilon_{k-2}^{(j+2)} - \epsilon_k^{(j)} \right| \tag{2.68}$$

is computed for every entry $\epsilon_k^{(j)}$ and the element at the bottom of the even column with the smallest $e_k^{(j)}$ is used as the extrapolated value I_i. A global error estimate is computed for the extrapolated I_i using

$$\varepsilon_i = |I_i - I_{i-1}| + |I_i - I_{i-2}| + |I_i - I_{i-3}| \tag{2.69}$$

where I_{i-1}, I_{i-2} and I_{i-3} are the previous three estimates of the global integral.

In 1983, Kahaner and Stoer [51] publish a similar algorithm which, in contrast to de Doncker's, uses the trapezoidal rule $T^{(m)}[a,b]$ as a basic rule and constructs a Romberg T-table [3] from the successive estimates, yet as with de Doncker's algorithm, it only sub-divides the intervals selectively and uses the last element $T_{i,i}$ of the T-table as the estimate I.

Kahaner and Stoer refer to de Doncker and note that both her use of a higher-order rule for $Q_n^{(m)}[a,b]$ and the use of the ϵ-Algorithm for extrapolation of both I and the error are much better choices.

Kahaner and Stoer suggest computing a conservative estimate of the error from the last three elements in the T-table:

$$\varepsilon = |T_{i,i} - T_{i,i-1}| + |T_{i,i} - T_{i-1,i-1}|. \tag{2.70}$$

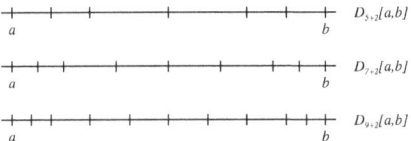

Figure 2.8: Ninomiya's error estimate approximates the $6th$, 8th and 10th derivative using 7, 9 and 11 points respectively.

No specific mention is made of how the error is approximated when using the ϵ-Algorithm, yet we may assume that again the difference between the final and the last two estimates is used. Note that the three last estimates are the last three elements in the *even* columns of the ϵ-table.

2.15 Ninomiya's Improved Adaptive Quadrature Method

In a 1980 paper [74], Ninomiya presents a recursive adaptive quadrature routine based on closed Newton-Cotes rules. He uses rules with $2n + 1$ nodes (results are given for 5, 7 and 9 points) and notes that these have an error of the form

$$\mathrm{NC}_{2n-1}[a, b] - \int_a^b f(x)\,\mathrm{d}x = K_{2n+1}(b - a)^{2n+1} f^{(2n)}(\xi), \quad \xi \in [a, b].$$

Instead of using the same quadrature rule on two or more sub-intervals to approximate the error as in Kuncir's and Lyness' error estimates, he adds two nodes in the center of the leftmost and the rightmost intervals. This choice is made

> *"taking account of the excellent properties of interpolation based on zeros of Chebyshev or Legendre polynomials which have high densities near the ends of [the] interval."*

Using $5 + 2$, $7 + 2$ and $9 + 2$ point stencils (Figure 2.8), he computes the error estimators

$$D_{5+2}[a, b] \approx \frac{(b-a)^7}{15\,120} f^{(6)}(\xi), \quad \xi \in [a, b] \tag{2.71}$$

$$D_{7+2}[a, b] \approx \frac{(b-a)^9}{1\,020\,600} f^{(8)}(\xi), \quad \xi \in [a, b] \tag{2.72}$$

$$D_{9+2}[a, b] \approx \frac{37(b-a)^{11}}{3\,066\,102\,400} f^{(10)}(\xi), \quad \xi \in [a, b], \tag{2.73}$$

which approximate the analytical error of the Newton-Cotes rules over 5, 7 and 9 points respectively. The weights of these estimators are pre-computed by solving a Vandermonde-like system of linear equations.

The error term is later subtracted from the initial estimate of the integral, resulting in a higher-degree estimate. The function values at the points added near the ends of the intervals are not used for the approximation of the integral itself. They can, however, be re-used if the interval needs to be sub-divided.

The resulting error estimate for the kth interval using $2n + 1$ points is approximated using

$$\varepsilon_k = |D_{2n+3}[a_k, b_k]| \frac{(b-a)}{(b_k - a_k)} \left(\log_2 \frac{(b-a)}{(b_k - a_k)} \right)^{-1}. \tag{2.74}$$

The interval-dependent scaling factor is proposed as an alternative to scaling by 2^d. Ninomiya states that

> "*Unfortunately, no adequate analytical or statistical theory concerning the application of the new method has been conceived as yet.*"

However, since bisection is used, the width of the kth interval can be written as

$$(b_k - a_k) = 2^{-d}(b - a) \tag{2.75}$$

where d is the depth of the kth subdivision (in a recursive scheme). Re-inserting this into Equation 2.74, we obtain

$$\varepsilon_k = |D_{2n+3}[a_k, b_k]| \frac{2^d}{d}, \tag{2.76}$$

which lies asymptotically below the 2^d first used by Kuncir (Section 2.2), yet larger than the $\sqrt{2}^d$ introduced by McKeeman (Section 2.4). This scaling is therefore also vulnerable to the problems described in Equation 2.8 and Equation 2.9.

2.16   Laurie's Sharper Error Estimate

In 1983, Laurie publishes a sharper error estimate [56] based on two quadrature rules $Q_\alpha[a,b]$ and $Q_\beta[a,b]$ of degree α and β respectively, where $\alpha > \beta$, or $\alpha = \beta$ and $Q_\alpha[a,b]$ is assumed to be more precise than $Q_\beta[a,b]$. Laurie states that whereas the difference of the two rules is generally a good approximation to the error of $Q_\beta[a,b]$, it is a severe overestimate of the error of $Q_\alpha[a,b]$. He therefore proposes, as an error estimate for the more precise $Q_\alpha[a,b]$, which is usually used as an approximation to the integral, the extrapolated error estimate

$$\varepsilon_k = \frac{\left(Q_\alpha^{(2)} - Q_\beta^{(2)}\right)\left(Q_\alpha^{(2)} - Q_\alpha^{(1)}\right)}{Q_\beta^{(2)} - Q_\beta^{(1)} - Q_\alpha^{(2)} + Q_\alpha^{(1)}} \tag{2.77}$$

where the ranges $[a_k, b_k]$ are omitted for simplicity. He shows that this error estimate is valid when

$$\left| Q_\alpha^{(2)} - Q_\alpha^{(1)} \right| < \left| Q_\beta^{(2)} - Q_\beta^{(1)} \right|, \tag{2.78}$$

which can be checked for in practice, and when

$$0 \leq \frac{Q_\alpha^{(2)} - I}{Q_\alpha^{(1)} - I} \leq \frac{Q_\beta^{(2)} - I}{Q_\beta^{(1)} - I} < 1 \tag{2.79}$$

which is impossible to verify since the exact integral $I = \int_a^b f(x)\,\mathrm{d}x$ must be known. These two conditions imply that the error of $Q_\alpha[a,b]$ is smaller than and decreases at a faster rate than that of $Q_\beta[a,b]$.

Laurie suggests a weaker condition that can be checked in practice, namely replacing I by $Q_\alpha^{(2)}[a_k, b_k] + \varepsilon_k$ in Equation 2.79, resulting in

$$0 \leq \frac{Q_\alpha^{(2)} - Q_\beta^{(2)}}{Q_\alpha^{(1)} - Q_\beta^{(1)}} < 1. \tag{2.80}$$

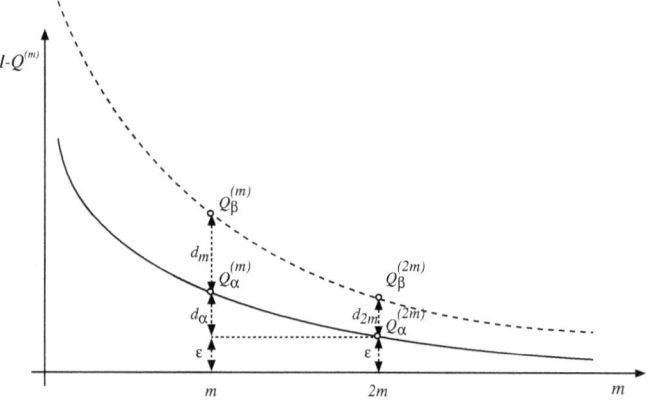

Figure 2.9: The error of the quadrature rules $Q_\alpha^{(m)}$ (solid curve) and $Q_\beta^{(m)}$ (dotted curve) as a function of the number of panels or subdivisions m.

Espelid and Sørevik [25] show, however, that this weaker condition is often satisfied when Equation 2.79 is not, which can lead to bad error estimates[15].

The error estimate itself, based on these assumptions, is best explained graphically (see Fig. 2.9). The errors of both rules $Q_\alpha^{(m)}[a, b]$ and $Q_\beta^{(m)}[a, b]$ are assumed to decrease exponentially with the increasing number of panels or subdivisions m:

$$Q_\alpha^{(m)} - I = \kappa_\alpha \left(\frac{b - a}{m}\right)^{\alpha+2} f^{(\alpha+1)}(\xi_\alpha), \quad \xi_\alpha \in [a, b], \quad (2.81)$$

$$Q_\beta^{(m)} - I = \kappa_\beta \left(\frac{b - a}{m}\right)^{\beta+2} f^{(\beta+1)}(\xi_\beta), \quad \xi_\beta \in [a, b], \quad (2.82)$$

where we will assume that $f^{(\alpha+1)}(x)$ and $f^{(\beta+1)}(x)$ are constant. Under this assumption, we can verify that Equation 2.79 holds for $\beta \leq \alpha$ by

---

[15]Espelid and Sørevik show that this is the case when using the 10-point Gauss rule and its 21-point Kronrod extension for $Q_\beta^{(1)}$ and $Q_\alpha^{(1)}$ respectively and integrating $\int_1^2 0.1/(0.01 + (x - \lambda)^2)\,dx$ for $1 \leq \lambda \leq 2$.

inserting Equation 2.81 and Equation 2.82 therein:

$$\frac{\kappa_\alpha \left(\frac{b-a}{2m}\right)^{\alpha+2} f^{(\alpha+1)}(\xi_\alpha)}{\kappa_\beta \left(\frac{b-a}{2m}\right)^{\beta+2} f^{(\beta+1)}(\xi_\beta)} \leq \frac{\kappa_\alpha \left(\frac{b-a}{m}\right)^{\alpha+2} f^{(\alpha+1)}(\xi_\alpha)}{\kappa_\beta \left(\frac{b-a}{m}\right)^{\beta+2} f^{(\beta+1)}(\xi_\beta)}$$

$$\frac{2^{-(\alpha+2)} \left(\frac{b-a}{m}\right)^{\alpha+2}}{2^{-(\beta+2)} \left(\frac{b-a}{m}\right)^{\beta+2}} \leq \frac{\left(\frac{b-a}{m}\right)^{\alpha+2}}{\left(\frac{b-a}{m}\right)^{\beta+2}}$$

$$\frac{2^{-(\alpha+2)}}{2^{-(\beta+2)}} \leq 1$$

$$2^{\beta-\alpha} \leq 1, \quad \beta \leq \alpha.$$

We then define the distances

$$d_\alpha = Q_\alpha^{(m)} - Q_\alpha^{(2m)} \tag{2.83}$$

$$d_m = Q_\beta^{(m)} - Q_\alpha^{(m)} \tag{2.84}$$

$$d_{2m} = Q_\beta^{(2m)} - Q_\alpha^{(2m)} \tag{2.85}$$

$$\varepsilon = Q_\alpha^{(2m)} - I \tag{2.86}$$

as they are shown in Figure 2.9. Using these differences, we can re-write the terms in Equation 2.79 as

$$Q_\alpha^{(2m)} - I = \varepsilon \tag{2.87}$$

$$Q_\beta^{(2m)} - I = \varepsilon + d_{2m} \tag{2.88}$$

$$Q_\alpha^{(m)} - I = \varepsilon + d_\alpha \tag{2.89}$$

$$Q_\beta^{(m)} - I = \varepsilon + d_\alpha + d_m. \tag{2.90}$$

Re-inserting these terms into Equation 2.79, we obtain

$$\frac{\varepsilon}{\varepsilon + d_\alpha} \geq \frac{\varepsilon + d_{2m}}{\varepsilon + d_\alpha + d_m}$$

$$\varepsilon(\varepsilon + d_\alpha + d_m) \geq (\varepsilon + d_{2m})(\varepsilon + d_\alpha)$$

$$\varepsilon(d_m + d_\alpha) \geq d_\alpha d_{2m} + \varepsilon(d_\alpha + d_{2m})$$

$$\varepsilon \leq \frac{d_\alpha d_{2m}}{d_{2m} - d_m}. \tag{2.91}$$

Re-inserting Equation 2.83–Equation 2.85, we see that this bound is identical to the error estimate proposed by Laurie in Equation 2.77.

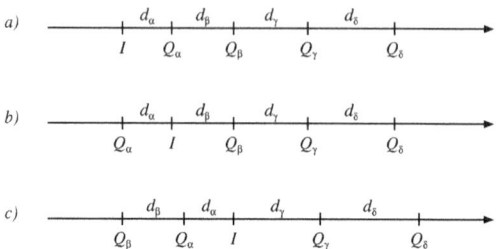

Figure 2.10: Different configurations of the four estimates Q_α, Q_β, Q_γ and Q_δ around the exact integral I.

This error estimate was tested both by Laurie himself in 1985 [57] and by Espelid and Sørevik in 1989 [25]. Laurie notes that while the error estimates are more accurate than other linear estimators, some reliability is lost, whereas Espelid and Sørevik show that when the error estimate is used in QUADPACK's QAG, the results can be catastrophic. In both cases, however, it is not specified if the conditions in Equation 2.78 and Equation 2.80 are tested for and if so, what is done if they fail.

In 1991, Favati, Lotti and Romani [27] publish a similar error estimator, based on four quadratures $Q_\alpha[a, b]$, $Q_\beta[a, b]$, $Q_\gamma[a, b]$ and $Q_\delta[a, b]$ of degree $\alpha > \beta > \gamma > \delta$ that satisfy

$$|I - Q_\alpha| \leq |I - Q_\delta| \tag{2.92}$$

$$|I - Q_\alpha| \leq |I - Q_\gamma| \tag{2.93}$$

$$|I - Q_\alpha| \leq |I - Q_\beta| \tag{2.94}$$

and the relation

$$\frac{|I - Q_\alpha|}{|I - Q_\gamma|} \leq \frac{|I - Q_\beta|}{|I - Q_\delta|}. \tag{2.95}$$

For any ordering of the four estimates Q_α, Q_β, Q_γ and Q_δ around the exact integral I, we can define the distances d_α, d_β, d_γ and d_δ. Note that the definition of the distances depends on the configuration of the estimates around I (see Fig. 2.10). The algorithm therefore first makes a decision as to which configuration is actually correct based on the differences between the actual estimates. Based on this decision, it computes the d_α, d_β, d_γ and d_δ or bounds them using Equation 2.92–

Equation 2.94 and inserts them into Equation 2.95 to extract an upper bound for $d_\alpha = |I - Q_\alpha|$, *i.e.* for the setup in Fig. 2.10a, and only for that setup, where $I < Q_\alpha < Q_\beta < Q_\gamma < Q_\delta$,

$$\frac{d_\alpha}{d_\alpha + d_\beta + d_\gamma} \leq \frac{d_\alpha + d_\beta}{d_\alpha + d_\beta + d_\gamma + d_\delta}$$

$$d_\alpha d_\delta \leq d_\alpha d_\beta + d_\beta^2 + d_\beta d_\gamma$$

$$d_\alpha \leq \frac{d_\beta^2 + d_\beta d_\gamma}{d_\delta - d_\beta} \qquad (2.96)$$

where

$$d_\beta = Q_\beta - Q_\alpha, \quad d_\gamma = Q_\gamma - Q_\beta, \quad d_\delta = Q_\delta - Q_\gamma.$$

Favati *et al.* test this algorithm on a number of integrands and show that the milder condition in Equation 2.95, which does not require successive estimates to decrease monotonically, is satisfied more often than that of Laurie in Equation 2.79.

2.17 QUADPACK's QAG Adaptive Quadrature Subroutine

In 1983, the most widely-used "commercial strength" quadrature subroutine library QUADPACK is published by Piessens *et al.* [83]. The general adaptive quadrature subroutine QAG is an extension of Piessens' integrator (see Section 2.12), yet with a slight modification to the local error estimate

$$\varepsilon_k = \tilde{I}_k \min\left\{ 1, \left(200 \frac{|\mathsf{G}_n[a_k, b_k] - \mathsf{K}_{2n+1}[a_k, b_k]|}{\tilde{I}_k} \right)^{3/2} \right\} \qquad (2.97)$$

where the default value of n is 10 and the value

$$\tilde{I}_k = \int_{a_k}^{b_k} \left| f(x) - \frac{\mathsf{K}_{2n+1}[a_k, b_k]}{b_k - a_k} \right| \, \mathrm{d}x,$$

which is also evaluated using the $\mathsf{K}_{2n+1}[a, b]$ rule, is used, as described by Krommer and Überhuber [53], as "*a measure for the smoothness of f on $[a, b]$*".

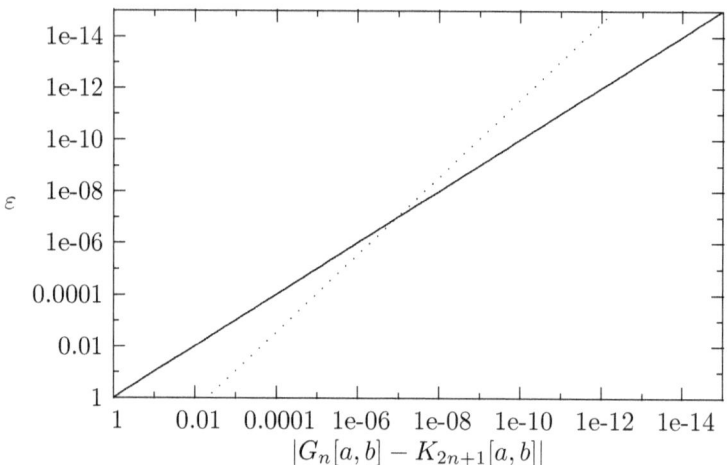

Figure 2.11: The error measure $(200\,|\mathsf{G}_n[a,b] - \mathsf{K}_{2n+1}[a,b]|)^{3/2}$ (dashed line) plotted as a function of $|\mathsf{G}_n[a,b] - \mathsf{K}_{2n+1}[a,b]|$.

As with Piessens' original error estimate (see Section 2.12), this estimate is also used in a non-recursive scheme as in Algorithm 2, hence the absence of a depth-dependent scaling factor and the problems involved therewith. The error measure

$$(200 \times |\mathsf{G}_n[a_k, b_k] - \mathsf{K}_{2n+1}[a_k, b_k]|)^{3/2}$$

is best explained graphically, as is done in Piessens *et al.* (Figure 2.11). The exponent $\frac{3}{2}$ is determined experimentally and scales the error exponentially, with a break-even point at 1.25×10^{-6} which is approximately relative machine precision for IEEE 754 32-bit floating point arithmetic. The scaling makes the estimate overly pessimistic for error estimates larger than 1.25×10^{-6} and overly optimistic for error estimates below that threshold. This measure is further divided by $\sqrt{\tilde{I}_k}$. Krommer and Überhuber explain this as follows:

"If this ratio is small, the difference between the two quadra-

ture formulas is small compared to the variation of f on $[a, b]$; i.e. , the discretization of f in the quadrature formulas G_n and K_{2n+1} is fine with respect to its variation. In this case, K_{2n+1} can indeed be expected to yield a better approximation for If than G_n."

Unfortunately, no further analysis is given in either [83] or [53].

Although difficult to interpret mathematically, this error estimate works well in practice. As Favati *et al.* [26] have pointed out, its tendency to over-estimate the error guard it from bad estimates on highly fluctuating, discontinuous or otherwise tricky integrands.

This local error estimate is re-used by Favati *et al.* [26], yet using pairs of "recursive monotone stable" (RMS) nested quadrature rules presented in [27], which allow for function evaluations to be re-used after bisection. The algorithm (using KEY=2 as described in [26]) is doubly-adaptive in the sense that it tries up to 4 rules of increasing degree before deciding to bisect the interval. They show this to be substantially more efficient than the original QAG's use of Gauss-Kronrod rules of a fixed order.

Hasegawa *et al.* [42] extend this approach by choosing bisection over increasing the degree of the quadrature rule when the ratio of two successive error estimates is larger than an empirically determined constant (as is suggested by Venter and Laurie in [99], see Section 2.10). They also use an interval-dependent scaling factor for the tolerance (their algorithm is recursive, not global as in QUADPACK's QAG) based on Ninomiya's quadrature method (see Section 2.15, Equation 2.74), replacing the $\log(\cdot)$ by a $\sqrt{\cdot}$:

$$\tau_k = \tau \frac{(b_k - a_k)}{(b - a)} \sqrt{\frac{(b - a)}{(b_k - a_k)}}$$

which, using the same analysis as in Equation 2.75, is equivalent to

$$\begin{aligned} \tau_k &= \tau \frac{2^{-d}(b - a)}{(b - a)} \sqrt{\frac{(b - a)}{2^{-d}(b - a)}} \\ &= \frac{\tau}{\sqrt{2}^d} \end{aligned} \tag{2.98}$$

which is the same as the local scaling introduced by McKeeman (see

Section 2.4) and hence also shares the same problems (see Equation 2.8 and Equation 2.9).

2.18   Berntsen and Espelid's Improved Error Estimate

In 1984 Berntsen and Espelid [4] suggest that instead of using the difference between a Gauss quadrature rule over n points and its Kronrod extension over $2n + 1$ points, one could directly use a Gauss quadrature rule over $2n+1$ points for the estimate of the integral. To estimate the error of this rule of degree $4n + 1$, they suggest removing one of the points and creating a new interpolatory quadrature rule $Q_{2n-1}[a, b]$ of degree $2n - 1$ over the remaining $2n$ points:

$$\varepsilon_k = \left| G_{2n+1}[a_k, b_k] - Q_{2n-1}[a_k, b_k] \right|. \tag{2.99}$$

Since the degree of the rule $Q_{2n-1}[a, b]$ is the same as that of the Gauss quadrature rule $G_n[a, b]$ used by Piessens (see Section 2.12), the error estimate is 0 for functions of up to the same algebraic degree of precision, yet the final estimate is n degrees higher: $4n+1$ for $G_{2n+1}[a, b]$ vs. $3n+1$ for $K_{2n+1}[a, b]$. A further advantage is the relative ease with which the weights of the rule $Q_{2n-1}[a, b]$ can be computed, as opposed to the effort required for the nodes and weights of the Kronrod extension.

2.19   Berntsen and Espelid's Null Rules

In a 1991 paper, Berntsen and Espelid [5] present an error estimator based on sequences of null rules.

Introduced by Lyness in 1965 [60], a null rule $N_n^{(k)}$ of degree k is defined as a set of weights $u_i^{(k)}$ over the n nodes x_i, $i = 1 \dots n$ such that

$$N_n^{(k)}[a, b] = \sum_{i=1}^{n} u_i^{(k)} x_i^j = \begin{cases} 0, & j \le k \\ \ne 0 & j = k+1 \end{cases}, \tag{2.100}$$

i.e. the rule evaluates all polynomials of degree $j \le k$ to 0 and the $(k+1)$st monomial to some non-zero value.

The null rules themselves can be constructed by solving a Vandermonde-like system of equations, which is under-determined for $k < n - 1$:

$$
\begin{pmatrix}
1 & 1 & \cdots & 1 \\
x_1 & x_2 & \cdots & x_n \\
x_1^2 & x_2^2 & \cdots & x_n^2 \\
\vdots & \vdots & \ddots & \vdots \\
x_1^{k+1} & x_2^{k+1} & \cdots & x_n^{k+1}
\end{pmatrix}
\begin{pmatrix}
u_1^{(k)} \\
u_2^{(k)} \\
\vdots \\
u_n^{(k)}
\end{pmatrix}
=
\begin{pmatrix}
0 \\
0 \\
0 \\
\vdots \\
1
\end{pmatrix}
\tag{2.101}
$$

Berntsen and Espelid compute a sequence of *orthogonal* null rules of decreasing degree. Two rules $N_n^{(j)}$ and $N_n^{(k)}$ are orthogonal if

$$
\sum_{i=1}^{n} u_i^{(j)} u_i^{(k)} = 0.
$$

The null rules are further normalized such that they are "equally strong", that is, given the weights w_i of the underlying interpolatory quadrature rule Q_n over the same nodes x_i, $i = 1 \ldots n$, for any null rule $N_n^{(k)}$,

$$
\sum_{i=1}^{n} \left(u_i^{(k)} \right)^2 = \sum_{i=1}^{n} w_i^2.
\tag{2.102}
$$

This set of orthogonal null rules $N_n^{(n-1)}$, $N_n^{(n-2)}$, ..., $N_n^{(0)}$ thus forms an orthogonal vector basis S_n in $\mathbb{R}^n$.

Applying the null rules to the integrand $f(x)$:

$$
e_k := N_n^{(k)}[a, b] = \sum_{i=1}^{n} u_i^{(k)} f(x_i), \quad k = 0 \ldots n - 1,
\tag{2.103}
$$

we obtain the *coefficients of the interpolation*[16] e_k of the integrand $f(x)$ onto S_n. Since the basis S_n is orthogonal, we can write

$$
f(x_i) = \frac{1}{\sum_{k=1}^{n} w_k^2} \sum_{k=0}^{n-1} e_k u_i^{(k)}, \quad i = 1 \ldots n.
\tag{2.104}
$$

---

[16]Let $p_k(x)$, $k = 0 \ldots n - 1$ be the polynomials of increasing degree k orthogonal with respect to the discrete scalar product

$$
(f, g) = \sum_{i=1}^{n} f(x_i) g(x_i),
$$

This connection between null rules, interpolation and orthogonal polynomials was already made by Laurie in [59], in which he also suggests different types of coefficients constructed using non-linear combinations of null rules.

The scaling of the null rules to match the "strength" of the quadrature rule, described in detail in [18], has no effect either on the basis or the interpolation itself. Any different scaling would result in a constant scaling of the coefficients e_k.

Following Berntsen and Espelid, to avoid "phase effects" as described in [63], the values are then paired as

$$E_k = \left(e_{2k}^2 + e_{2k+1}^2\right)^{1/2}, \quad k = 0 \ldots n/2 - 1. \tag{2.105}$$

and the ratio of these pairs is computed

$$r_k = \frac{E_k}{E_{k+1}}, \quad k = 0 \ldots n/2 - 2 \tag{2.106}$$

and the largest of the last K ratios

$$r_{\mathsf{max}} = \max_k r_k,$$

is taken as an estimate of the convergence rate of the coefficients. If this ratio is larger than 1, then the function is assumed to be "non-asymptotic" in the interval and the largest E_k is used as a local error estimate.

If r_{max} is below 1 yet still above some critical value r_{critical}, the function is assumed to be "weakly asymptotic" and the value of the next-highest coefficient $E_{n/2+1}$ — and thus the local error — is estimated using

$$\varepsilon_k = 10 r_{\mathsf{max}} E_{n/2-1} \tag{2.107}$$

---

that interpolate the weights of the respective null rules, such that

$$p_k(x_i) = u_i^{(k)}, \quad i = 1 \ldots n$$

then the coefficients e_k computed in Equation 2.103 are also the coefficients of interpolation of the integrand at the nodes of the quadrature rule such that

$$f(x_i) = \sum_{k=0}^{n-1} e_k p_k(x_i), \quad i = 1 \ldots n.$$

Finally, if r_{max} is below the critical ratio, then the function is assumed to be *"strongly asymptotic"* and the error is estimated using

$$\varepsilon_k = 10 r_{critical}^{1-\alpha} r_{max}^{\alpha} E_{n/2-1}. \tag{2.108}$$

where $\alpha \geq 1$ is chosen to reflect, as Berntsen and Espelid state, *"the degree of optimism we want to put into this algorithm."*

In summary we can say that the coefficients of the polynomial interpolating the integrand at the nodes of the quadrature rule, relative to some discretely orthogonal basis, are computed (Equation 2.103) and paired (Equation 2.105). The monotonic decrease of these pairs ($r_{max} < 1$) is used as an indicator for the asymptoticity of the function in the interval. If the function is deemed sufficiently asymptotic, the ratio of decrease of the coefficient pairs is used to extrapolate the magnitude of the next coefficient pair, which is used as an error estimate.

Berntsen and Espelid implement and test this error estimate using, as basic quadrature rules, 21-point Gauss, Lobatto, Gauss-Kronrod and Clenshaw-Curtis rules as well as 61-point Gauss and Gauss-Kronrod rules, and later in DQAINT [19], based on QUADPACK's QAG (see Section 2.17), using as a basic quadrature rule the Gauss, Gauss-Lobatto and Gauss-Kronrod rules over 21 nodes with the constants

$$r_{critical} = 1/4 \tag{2.109}$$
$$\alpha = (d - (n - 4))/4 \tag{2.110}$$
$$K = 3, \tag{2.111}$$

where d is the degree of the underlying quadrature rule.

In [20] and [21], Espelid implements this error estimate in four adaptive quadrature routines based on Newton-Cotes rules. The first of these routines, modsim, is an adaptation of Gander and Gautschi's recursive adaptsim (see Section 2.20) and uses the 5-point Newton-Cotes rule $NC_5^{(1)}[a, b]$ over which the last 4 null rules are computed. Due to the small number of points, the null rules are not paired as in Equation 2.105 and their ratios are computed directly. As with adaptsim, the global tolerance is also used as the local tolerance (*i.e.* the error estimate is not scaled by a depth-dependent factor), meaning that for a large number of intervals, the global error may be above the required tolerance, even if the tolerance is satisfied in all local intervals.

The second routine, `modlob`, differs only in that it uses the 7-point Kronrod extension of a 4-point Gauss-Lobatto rule used by Gander and Gautschi's `adaptlob` [34] (see Section 2.20). Here too only the last 4 null rules are computed and the resulting values are not paired.

The third routine, `coteda`, uses what Espelid terms a *"doubly adaptive"*[17] strategy of first applying a 5-point Newton-Cotes rule and estimating the error as in `modsim` above and, if the requested tolerance is not satisfied, extending it to a 9-point Newton-Cotes rule. The error of the 9-point rule is computed using the last 8 null rules and by pairing the coefficients as in Equation 2.105. If this second error estimate is above the requested tolerance, then the interval is subdivided. Here too the global tolerance is used in every subinterval.

The fourth and last routine, `coteglob`, is a global (as in Algorithm 2) version of `coteda` and was added to illustrate the difference between local and global strategies, reinforcing the conclusions by Malcolm and Simpson [66].

In [22] and [24] this global doubly adaptive strategy is extended to four Newton-Cotes rules over 5, 9, 17 and 33 equidistant points. The error estimates for the 5 and 9 point rules are computed as above and for the 17 and 33 point rules, 15 null rules are computed and grouped into triplets instead of pairs:

$$E_k = \left(e_{3k}^2 + e_{3k+1}^2 + e_{3k+2}^2\right)^{1/2}, k = 0 \ldots (n-1)/3.$$

As opposed to the previous doubly adaptive integrators, the interval is subdivided if, for a rule over $2n - 1$ points, the error estimate over that rule is larger than the sum of the errors of the n point rule over the left and right half-intervals.

Some details must be observed regarding the order of the null rules used: For the 5 and 9 point stencils, the normal Newton-Cotes rules are used,

---

[17]The term "doubly adaptive" may be somewhat abusive in this context since, in fact, only two successive quadrature rules are applied and the decision to bisect is only taken *after* the highest-order rule has failed, as opposed to Oliver's doubly-adaptive quadrature (Section 2.11), where, for every interval, the decision to subdivide *or* increase the degree of the quadrature rule is evaluated. Interestingly enough, Espelid credits Cools and Haegmans [11] with having published the first doubly-adaptive strategy, although they were preceded by de Boor in 1971 (Section 2.9), Oliver in 1972 (Section 2.11) and Patterson in 1973 (Section 2.12) — an error he corrects (somewhat) in [24] by attributing the first publication to Oliver in 1972.

yet for the 17 and 33 point stencils, the Newton-Cotes rules are numerically unstable. Therefore, rules of degree 13, 15 and 17 (for the 17 point stencil) and 21, 23 and 27 (for the 33 point stencil) are constructed such as to minimize truncation errors (*i.e.* the sum of squares of the quadrature weights). The degrees of these rules, however, are well below the degree of the moments that the null rules approximate over the same points. Espelid treats this problem by choosing α *"to fit the order of the actual rule used, the order of $r_{\max}$ and of $E_{n/3+2}$."*

In [23], Espelid compares these routines to QUADPACK's QAG and concludes that they are generally more efficient.

2.20 Gander and Gautschi's Revisited Adaptive Quadrature

In a 2001 paper, Gander and Gautschi [34] present two recursive adaptive quadrature routines. The first of these routines, adaptsim is quite similar to Lyness' SQUANK (see Section 2.7). It computes the approximations $S^{(1)}[a, b]$ and $S^{(2)}[a, b]$ and uses them to extrapolate $NC_5^{(1)}[a, b]$ as in Equation 2.19. The *globally relative* local error estimate, however, is then computed as

$$\varepsilon_k = \left| NC_5^{(1)}[a_k, b_k] - S^{(2)}[a_k, b_k] \right| / |\hat{I}| \tag{2.112}$$

where $\hat{I}$ is a rough approximation to the global integral computed over a set of random nodes (if this approximation yields 0, then $b - a$ is used instead to avoid division by zero). If the tolerance is met, the approximation $NC_5^{(1)}[a_k, b_k]$ is used for the integral, otherwise, the interval is bisected, re-using all evaluations of the integrand.

To guard against machine-dependent problems and to prevent the algorithm from trying to achieve spurious accuracy, *i.e.* when the local interval is less than $\varepsilon_{\mathsf{mach}}\hat{I}$, the termination criterion

$$\frac{\tau}{\varepsilon_{\mathsf{mach}}}\hat{I} + \left(NC_5^{(1)}[a_k, b_k] - S^{(2)}[a_k, b_k] \right) = \frac{\tau}{\varepsilon_{\mathsf{mach}}}\hat{I} \tag{2.113}$$

where $\varepsilon_{\mathsf{mach}}$ is the machine precision, is used. This termination criterion exploits the fact that if the error estimate is below tolerance, it will be canceled-out in the addition.

In order to avoid cases in which the error estimate will not converge (as, for example, Equation 2.8), the algorithm computes the midpoint of the interval $m = (a_k + b_k)/2$ and checks if

$$(a_k < m) \land (m < b_k).$$

If this condition is not satisfied then there is no machine number between a and b and further bisection is neither reasonable nor possible. In such a case, the recursion is aborted and an error message is issued that the requested tolerance may not have been met.

The second routine, `adaptlob`, uses a 4-point Gauss-Lobatto rule $\mathsf{GL}_4^{(1)}[a,b]$ and its 7-point Kronrod extension $\mathsf{K}_7^{(1)}[a,b]$. The globally relative local error is computed, analogously to Equation 2.112, as

$$\varepsilon_k = \left| \mathsf{GL}_4^{(1)}[a_k, b_k] - \mathsf{K}_7^{(1)}[a_k, b_k] \right| / |\hat{I}| \qquad (2.114)$$

and evaluated using the same scheme as in Equation 2.113. If the tolerance is met, the approximation $\mathsf{K}_7^{(1)}[a,b]$ is used for the integral. Otherwise, the interval is divided into the 6 sub-intervals between the nodes of the Kronrod extension, thus re-using all evaluations of the integrand.

Note that although both algorithms are recursive as per Algorithm 1, the error estimates are not subject to a depth-dependent scaling factor. The error estimates of both algorithms thus suffer from a common problem already discussed in the context of Patterson's integrator (see Section 2.12), namely that the sum of the local error estimates, which are each below the required tolerance, is not guaranteed to be itself below the required tolerance. This, however, also avoids the problems related to using such a scaling factor, as seen in Equation 2.8 and Equation 2.9.

These two algorithms, with modifications (*i.e.* a cap on the number of function evaluations and on the recursion depth and an absolute tolerance requirement), have been added to MATLAB [67] as of release 6.5 as the standard integrators `quad` and `quadl`.

Chapter 3

Function Representation

3.1 Function Representation using Orthogonal Polynomials

In most quadrature algorithms, the integrand is not explicitly represented, except through different approximations $Q_n^{(m)}$ of its integral. The *differences* between these different approximations are used to extrapolate information on the integrand, such as higher derivatives, used to approximate the quadrature error.

Some authors such as Gallaher (see Section 2.6) or Ninomiya (see Section 2.15) use additional nodes to numerically approximate the higher derivative directly using divided differences, thus supplying additional information on the integrand $f(x)$.

Other authors such as O'Hara and Smith (see Section 2.8), Oliver (see Section 2.11) and Berntsen and Espelid (see Section 2.19) compute some of the higher-order coefficients of the function relative to some orthogonal base, thus further characterizing the integrand.

In all these cases, however, the characterization is not complete and in most cases only implicit. In the following, we will attempt to characterize the integrand as completely as possible.

Before further characterizing the integrand, we note that for every interpolatory quadrature rule, we are in fact computing an interpolation polynomial $g_n(x)$ of degree n such that

$$g_n(x_i) = f(x_i), \quad i = 0 \ldots n,$$

and evaluating the integral thereof

$$Q_n[a, b] = \int_a^b g_n(x) \, dx.$$

This equivalence is easily demonstrated, as is done in many textbooks in numerical analysis ([94, 90, 38, 91, 84] to name a few), by constructing the interpolation $g_n(x)$ using Lagrange polynomials

$$g_n(x) = \sum_{i=0}^{n} \ell_i(x) f(x_i)$$

and integrating $g_n(x)$

$$
\begin{aligned}
\int_a^b g_n(x) \, dx &= \int_a^b \sum_{i=0}^{n} \ell_i(x) f(x_i) \, dx \\
&= \sum_{i=0}^{n} f(x_i) \int_a^b \ell_i(x) \, dx
\end{aligned}
$$

to obtain the weights of the quadrature rule from Equation 1.1 with

$$w_i = \int_a^b \ell_i(x) \, dx.$$

Since any polynomial interpolation of degree n over $n + 1$ distinct points is uniquely determined, it does not matter how we represent $g_n(x)$ – its integral will always be identical to the result of the interpolatory quadrature rule $Q_n[a, b]$ over the same nodes.

In the following, we will represent $g_n(x)$ as a weighted sum of orthogonal polynomials:

$$g_n(x) = \sum_{i=0}^{n} c_i p_i(x) \qquad (3.1)$$

where the $p_i(x)$, $i = 0 \ldots n$ are polynomials of degree i which are orthonormal with respect to some product

$$(p_i, p_j) = \begin{cases} 0 & i \neq j, \\ 1 & i = j. \end{cases}$$

We will use the coefficients $\mathbf{c} = (c_0, c_1, \ldots, c_n)^\mathsf{T}$ as our representation of $g_n(x)$.

Furthermore, for notational simplicity, we will assume that the integrand has been transformed from the interval $[a, b]$ to the interval $[-1, 1]$. The polynomial $g_n(x)$ interpolates the integrand $f(x)$ at the nodes $x_i \in [-1, 1]$:

$$g_n(x_i) = f(x_i), \quad i = 0 \ldots n.$$

In the following, we will use the orthonormal Legendre polynomials, which are orthogonal with respect to the product

$$(p_i, p_j) = \int_{-1}^{1} p_i(x) p_j(x) \, \mathrm{d}x,$$

and satisfy the three-term recurrence relation

$$\alpha_k p_{k+1}(x) = (x + \beta_k) p_k(x) - \gamma_k p_{k-1}(x)$$

with the coefficients

$$\alpha_k = \sqrt{\frac{(k+1)^2}{(2k+1)(2k+3)}}, \quad \beta_k = 0, \quad \gamma_0 = \sqrt{2}, \quad \gamma_k = \sqrt{\frac{k^2}{4k^2 - 1}}.$$

Note that although we will use the orthonormal Legendre polynomials, the following derivations will still be valid for any set of polynomials orthonormal with respect to the product

$$(p_i, p_j) = \int_{-1}^{1} p_i(x) p_j(x) w(x) \, \mathrm{d}x \tag{3.2}$$

for any measure $w(x)$ positive on $[-1, 1]$. The resulting norms will, however, also all be defined with respect to the measure $w(x)$.

The representation of $g_n(x)$ using orthogonal polynomials (Equation 3.1) has some interesting properties. First of all, it is simple to evaluate the integral of $g_n(x)$ using

$$
\begin{aligned}
\int_{-1}^{1} g_n(x)\,\mathrm{d}x &= \int_{-1}^{1} \sum_{i=0}^{n} c_i p_i(x)\,\mathrm{d}x \\
&= \sum_{i=0}^{n} c_i \underbrace{\int_{-1}^{1} p_i(x)\,\mathrm{d}x}_{=\omega_i} \\
&= \boldsymbol{\omega}^{\mathsf{T}} \mathbf{c}
\end{aligned}
$$

where the weights $\boldsymbol{\omega}^{\mathsf{T}}$ can be pre-computed and applied much in the same way as the weights of a quadrature rule. Note that for the normalized Legendre polynomials used herein, $\boldsymbol{\omega}^{\mathsf{T}} = (1/\sqrt{2}, 0, \ldots, 0)$.

We can also evaluate the L_2-norm of $g_n(x)$ quite efficiently using Parseval's theorem:

$$
\left[\int_{-1}^{1} g_n^2(x)\,\mathrm{d}x \right]^{1/2} = \left[\sum_{i=0}^{n} c_i^2 \right]^{1/2} = \|\mathbf{c}\|_2.
$$

In the following, we will use $\|\cdot\|$ to denote the 2-norm[1].

A final useful feature is that, given the coefficients of $g_n(x)$ on $[a, b]$, we can construct upper-triangular matrices

$$
T_{i,j}^{(\ell)} = \int_{-1}^{1} p_i(x) p_j((x-1)/2)\,\mathrm{d}x, \quad i = 0 \ldots n, \; j \geq i
$$

and

$$
T_{i,j}^{(r)} = \int_{-1}^{1} p_i(x) p_j((x+1)/2)\,\mathrm{d}x, \quad i = 0 \ldots n, \; j \geq i
$$

such that

$$
\mathbf{c}^{(\ell)} = \mathbf{T}^{(\ell)} \mathbf{c} \quad \text{and} \quad \mathbf{c}^{(r)} = \mathbf{T}^{(r)} \mathbf{c} \tag{3.3}
$$

are the coefficients of $g_n(x)$ on the intervals $[a, \frac{a+b}{2}]$ and $[\frac{a+b}{2}, b]$ respectively. If $\mathbf{c}$ are the coefficients of any given polynomial $g_n(x)$ on the interval $[a, b]$, then $\mathbf{c}^{(\ell)}$, as computed in Equation 3.3, are the coefficients of a

---

[1]Note that if we use different orthonormal polynomials with respect to some measure $w(x)$, we would obtain the L_2-norm with respect to that measure.

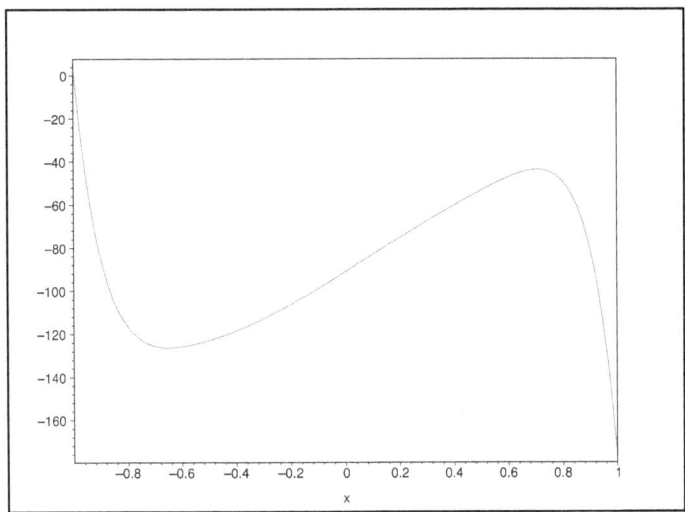

Figure 3.1: The interpolation polynomial $g_n(x)$ (red) and the polynomial $g_n^{(\ell)}(x)$ resulting from shifting its coefficients to the left half-interval as per Equation 3.3 (green). The polynomial $g_n^{(\ell)}(x)$ matches the left half of $g_n(x)$.

polynomial $g_n^{(\ell)}(x)$ matching $g_n(x)$ on the left half-interval $[a, (a+b)/2]$ such that

$$g_n^{(\ell)}(x) = g_n\left(\frac{x-1}{2}\right), \quad x \in [-1, 1]$$

as is shown in Figure 3.1. The coefficients $\mathbf{c}^{(\ell)}$ can be useful if, after bisecting an interval, we want to re-use, inside one of the sub-intervals, the interpolation computed over the entire original interval.

These matrices depend only on the polynomials $p_i(x)$ and can therefore be pre-computed for any set of nodes.

3.2   Interpolation Algorithms

In many applications, we need to compute polynomial interpolations of data or functions. In practical terms, this means that given $n+1$ function values f_i at the $n+1$ nodes x_i, $i = 0 \ldots n$, we want to construct a polynomial $g_n(x)$ of degree n such that

$$g_n(x_i) = f_i, \quad i = 0 \ldots n. \tag{3.4}$$

That is, the polynomial $g_n(x)$ interpolates the $n+1$ function values f_i at the nodes x_i.

There are many ways in which $g_n(x)$ can be represented. The most common thereof, which is often used as an introduction to interpolation in many textbooks ([94, 90, 38, 91, 84] to name a few), is the interpolation using Lagrange polynomials:

$$g_n(x) = \sum_{i=0}^{n} f_i \ell_i(x) \tag{3.5}$$

where the Lagrange polynomials are defined as

$$\ell_i(x) = \prod_{\substack{j=0 \\ j \neq i}}^{n} \frac{x - x_j}{x_i - x_j}. \tag{3.6}$$

These polynomials satisfy, by construction,

$$\ell_i(x_j) = \begin{cases} 0 & i \neq j \\ 1 & i = j \end{cases} \tag{3.7}$$

such that the condition in Equation 3.4 is also satisfied.

Although the representation with Lagrange polynomials is usually only considered as a didactical example, since its definition is also an algorithm for its construction, a more useful and computationally efficient variant thereof called *barycentric interpolation* [6] is still used in practical applications (*e.g.* see [2]).

Another representation constructs $g_n(x)$ using the Newton polynomials

$$\pi_i(x) = \prod_{j=0}^{i-1} (x - x_j). \tag{3.8}$$

The coefficients a_i of the interpolation

$$g_n(x) = \sum_{i=0}^{n} a_i \pi_i(x) \tag{3.9}$$

are computed from the divided differences

$$a_i = f[x_0, \ldots, x_i]. \tag{3.10}$$

As with the Lagrange interpolation, Equation 3.4 is satisfied by construction. One advantage of this representation is that it can be easily
extended, *i.e.* additional nodes and function values added, since, unlike
the Lagrange interpolation, the first k polynomials and coefficients depend only on the first k nodes and function values.

Yet another common representation for $g_n(x)$ uses monomials:

$$g_n(x) = \sum_{i=0}^{n} b_i x^i. \tag{3.11}$$

Unlike the representation with Lagrange or Newton polynomials, the interpolation property is not guaranteed by construction and the coefficients b_i must be computed such that Equation 3.4 is satisfied. These
constraints themselves form a system of $n+1$ linear equations which can
be solved for the $n+1$ coefficients b_i

$$\begin{pmatrix} 1 & x_0 & x_0^2 & \cdots & x_0^n \\ 1 & x_1 & x_1^2 & \cdots & x_1^n \\ \vdots & \vdots & \vdots & \ddots & \vdots \\ 1 & x_n & x_n^2 & \cdots & x_n^n \end{pmatrix} \begin{pmatrix} b_0 \\ b_1 \\ \vdots \\ b_n \end{pmatrix} = \begin{pmatrix} f_0 \\ f_1 \\ \vdots \\ f_n \end{pmatrix}. \tag{3.12}$$

The matrix in Equation 3.12 is called the *Vandermonde* matrix and the
system of equations can be solved naively using Gaussian elimination at
a cost in $\mathcal{O}(n^3)$, which is not especially economical. Furthermore, as
Gautschi shows in [36], for some common node distributions, the condition number of the Vandermonde matrix grows exponentially with the
matrix dimension, making a precise numerical computation of the coefficients impossible already for moderate n.

A final representation, which will be the main focus of this paper, is the
linear combination of polynomials

$$g_n(x) = \sum_{i=0}^{n} c_i p_i(x) \tag{3.13}$$

where the polynomials $p_i(x)$ of degree i can be constructed using a three-term recurrence relation, which we will write as[2]

$$\alpha_k p_{k+1}(x) = x p_k(x) + \beta_k p_k(x) - \gamma_k p_{k-1}(x), \qquad (3.14)$$

with

$$p_0(x) = 1, \quad p_{-1}(x) = 0.$$

Examples of such polynomials are the Legendre polynomials $P_k(x)$ with

$$\alpha_k = \frac{k}{2k-1}, \quad \beta_k = 0, \quad \gamma_k = \frac{k-1}{2k-1} \qquad (3.15)$$

which are orthogonal with respect to the scalar product

$$(f, g) = \int_{-1}^{1} f(x) g(x) \, dx$$

or the Chebyshev polynomials $T_k(x)$ with

$$\alpha_0 = 1, \quad \alpha_k = \frac{1}{2}, \quad \beta_k = 0, \quad \gamma_k = \frac{1}{2} \qquad (3.16)$$

which satisfy the trigonometric identity

$$T_k(\cos \theta) = \cos(k\theta)$$

in $[-1, 1]$ and whose properties, such as the minimal infinity-norm, have led to their wide use in interpolation and approximation applications [2, 86].

As with the coefficients of the monomials, the coefficients c_i of Equation 3.13 can be computed solving the system of linear equations

$$\begin{pmatrix} p_0(x_0) & p_1(x_0) & \cdots & p_n(x_0) \\ p_0(x_1) & p_1(x_1) & \cdots & p_n(x_1) \\ \vdots & \vdots & \ddots & \vdots \\ p_0(x_n) & p_1(x_n) & \cdots & p_n(x_n) \end{pmatrix} \begin{pmatrix} c_0 \\ c_1 \\ \vdots \\ c_n \end{pmatrix} = \begin{pmatrix} f_0 \\ f_1 \\ \vdots \\ f_n \end{pmatrix} \qquad (3.17)$$

---

[2]Gautschi [39] and Higham [46] use different recurrence relations which can both be transformed to the representation used herein.

which can be written as

$$\mathbf{P}^{(n)}\mathbf{c}^{(n)} = \mathbf{f}^{(n)}. \qquad (3.18)$$

The matrix $\mathbf{P}^{(n)}$ is a *Vandermonde-like* matrix and the system of equations can be solved naively in $\mathcal{O}(n^3)$ using Gaussian elimination. As with the computation of the monomial coefficients, the matrix may be ill-conditioned [37].

Note that the previously described Newton polynomials also satisfy a three-term recurrence relation with

$$\alpha_k = 1, \quad \beta_k = -x_{k-1}, \quad \gamma_k = 0$$

as do the monomials, with

$$\alpha_k = 1, \quad \beta_k = 0, \quad \gamma_k = 0. \qquad (3.19)$$

Creating the matrix $\mathbf{P}^{(n)}$ using the Newton polynomials results in a lower-triangular matrix since, by the definition in Equation 3.8, $\pi_k(x_i) = 0$ for all $i < k$. Solving this lower-triangular system using back-substitution results in the divided differences for the coefficients d_i (see Equation 3.10). This computation is straight-forward and can be done in $\mathcal{O}(n^2)$, which is a further advantage of using Newton polynomials.

If the nodes x_i are distinct then the polynomial that interpolates the f_i at these nodes is unique and all these representations – Equation 3.5, Equation 3.9, Equation 3.11 and Equation 3.13 – are different representations of the same polynomial. It therefore seems intuitive that it should be possible to convert the coefficients of one representation to those of another. In [31], such a connection is made and a set of elegant basis transformations for moving between Lagrange, monomial, Newton and (discrete) orthogonal polynomial interpolations is given.

This was already done implicitly by Björck and Pereyra in [7], where an algorithm is presented to compute the monomial coefficients of an interpolation, without having to solve the expensive and unstable Vandermonde system using Gaussian elimination, by computing first the coefficients of a Newton interpolation and then converting these to monomial coefficients. This approach was later extended by Higham in [46] to compute the coefficients for any polynomial basis satisfying a three-term recurrence relation.

In Section 3.2.1, we will re-formulate the algorithms of Björck and Pereyra and of Higham and extend them to facilitate the *downdate* of an interpolation, *i.e.* the removal of a node. In Section 3.2.2 we present a new algorithm for the construction of interpolations of the type of Equation 3.13 based on a successive decomposition of the Vandermonde-like matrix in Equation 3.17. Finally, in Section 3.2.3, we will present results regarding the efficiency and stability of the two algorithms.

3.2.1 A Modification of Björck and Pereyra's and of Higham's Algorithms Allowing Downdates

In [7], Björck and Pereyra present an algorithm which exploits the recursive definition of the Newton polynomials

$$\pi_k(x) = (x - x_{k-1})\pi_{k-1}(x). \tag{3.20}$$

They note that given the Newton interpolation coefficients a_i, the interpolation polynomial can be constructed using Horner's scheme:

$$q_n(x) = a_n, \quad q_k(x) = (x - x_k)q_{k+1}(x) + a_k, \quad k = n - 1 \ldots 0 \tag{3.21}$$

where the interpolation polynomial is $g_n(x) = q_0(x)$.

They also note that given a monomial representation for $q_k(x)$, such as

$$q_k(x) = \sum_{i=0}^{n-k} b_i^{(k)} x^i$$

the polynomial $q_{k-1}(x)$ can then be constructed, following the recursion in Equation 3.21, as

$$
\begin{aligned}
q_{k-1}(x) &= (x - x_{k-1})q_k(x) + a_{k-1} \\
&= (x - x_{k-1}) \sum_{i=0}^{n-k} b_i^{(k)} x^i + a_{k-1} \\
&= \sum_{i=1}^{n-k+1} b_{i-1}^{(k)} x^i - x_{k-1} \sum_{i=0}^{n-k} b_i^{(k)} x^i + a_{k-1} \\
&= b_{n-k}^{(k)} x^{n-k+1} + \sum_{i=1}^{n-k} (b_{i-1}^{(k)} - x_{k-1} b_i^{(k)}) x^i \\
&\quad + b_0^{(k-1)} + a_{k-1}.
\end{aligned}
\tag{3.22}
$$

From Equation 3.22 we can then extract the new coefficients $b_i^{(k-1)}$:

$$
b_i^{(k-1)} = \begin{cases}
b_{i-1}^{(k)} & i = n - k + 1 \\
b_{i-1}^{(k)} - x_{k-1} b_i^{(k)}, & 1 \le i \le n - k \\
b_0^{(k)} + a_{k-1}, & i = 0.
\end{cases}
\tag{3.23}
$$

In [46], Higham uses the same approach, yet represents the Newton polynomials as a linear combination of polynomials satisfying a three-term recurrence relation as shown in Equation 3.13. Using such a representation

$$
q_k(x) = \sum_{i=0}^{n-k} c_i^{(k)} p_i(x)
\tag{3.24}
$$

he computes $q_{k-1}(x)$ by expanding the recursion in Equation 3.21 using the representation Equation 3.24 and the recurrence relation in Equation 3.14:

$$
\begin{aligned}
q_{k-1}(x) &= (x - x_{k-1}) \sum_{i=0}^{n-k} c_i^{(k)} p_i(x) + a_{k-1} \\
&= \sum_{i=0}^{n-k} c_i^{(k)} x p_i(x) - x_{k-1} \sum_{i=0}^{n-k} c_i^{(k)} p_i(x) + a_{k-1} \\
&= \sum_{i=0}^{n-k} c_i^{(k)} \left(\alpha_i p_{i+1}(x) - \beta_i p_i(x) + \gamma_i p_{i-1}(x) \right) \\
&\quad -x_{k-1} \sum_{i=0}^{n-k} c_i^{(k)} p_i(x) + a_{k-1}.
\end{aligned}
\tag{3.25}
$$

Expanding Equation 3.25 for the individual $p_k(x)$, and keeping in mind that $p_{-1}(x) = 0$, we obtain

$$
\begin{aligned}
q_{k-1}(x) &= \sum_{i=1}^{n-k+1} c_{i-1}^{(k)} \alpha_{i-1} p_i(x) - \sum_{i=0}^{n-k} c_i^{(k)} \left(x_{k-1} + \beta_i \right) p_i(x) \\
&\quad + \sum_{i=0}^{n-k-1} c_{i+1}^{(k)} \gamma_{i+1} p_i(x) + a_{k-1}.
\end{aligned}
\tag{3.26}
$$

By shifting the sums in Equation 3.26 and re-grouping around the individual $p_k(x)$ we finally obtain

$$
\begin{aligned}
q_{k-1}(x) &= c_{n-k}^{(k)} \alpha_{n-k} p_{n-k+1}(x) \\
&\quad + \left(c_{n-k-1}^{(k)} \alpha_{n-k-1} - c_{n-k}^{(k)} (x_{k-1} + \beta_{n-k}) \right) p_{n-k}(x) \\
&\quad + \sum_{i=1}^{n-k-1} \left(c_{i-1}^{(k)} \alpha_{i-1} - c_i^{(k)} (x_{k-1} + \beta_i) + c_{i+1}^{(k)} \gamma_{i+1} \right) p_i(x) \\
&\quad - c_0^{(k)} (x_{k-1} + \beta_0) + c_1^{(k)} \gamma_1 + a_{k-1}.
\end{aligned}
\tag{3.27}
$$

Higham then extracts the new coefficients $c_i^{(k-1)}$ from Equation 3.27 as:

$$
c_i^{(k-1)} = \begin{cases}
c_{i-1}^{(k)}\alpha_{i-1}, & i = n-k+1 \\
c_{0-1}^{(k)}\alpha_{i-1} - c_i^{(k)}(x_{k-1}+\beta_i), & i = n-k \\
c_{i-1}^{(k)}\alpha_{i-1} - c_i^{(k)}(x_{k-1}+\beta_i) + c_{i+1}^{(k)}\gamma_{i+1}, & 1 \le i < n-k \\
-c_0^{(k)}(x_{k-1}+\beta_0) + c_1^{(k)}\gamma_1 + a_{k-1} & i = 0
\end{cases}
$$
(3.28)

In both algorithms, the interpolating polynomial is constructed by first computing the a_i, $i = 0 \ldots n$ and, starting with $q_n(x) = a_n$, and hence $c_0^{(n)} = a_n$ or $b_0^{(n)} = a_n$, and successively updating the coefficients per Equation 3.28 and Equation 3.23.

Alternatively, we could use the same approach to compute the coefficients of the Newton polynomials themselves

$$
\pi_k(x) = \sum_{i=0}^{k} \eta_i^{(k)} p_i(x).
$$

Expanding the recurrence relation in Equation 3.20 analogously to Equation 3.27, we obtain

$$
\begin{aligned}
\pi_{k+1}(x) &= \eta_k^{(k)}\alpha_k p_{k+1}(x) + \left(\eta_{k-1}^{(k)}\alpha_{k-1} - \eta_k^{(k)}(x_k+\beta_k) \right) p_k(x) \\
&+ \sum_{i=1}^{k-1} \left(\eta_{i-1}^{(k)}\alpha_{i-1} - \eta_i^{(k)}(x_k+\beta_i) + \eta_{i+1}^{(k)}\gamma_{i+1} \right) p_i(x) \\
&- \eta_0^{(k)}(x_k+\beta_0) + \eta_1^{(k)}\gamma_1.
\end{aligned}
$$
(3.29)

We initialize with $\eta_0^{(0)} = 1$ and use

$$
\eta_i^{(k+1)} = \begin{cases}
\eta_{i-1}^{(k)}\alpha_{i-1}, & i = k+1 \\
\eta_{i-1}^{(k)}\alpha_{i-1} - \eta_i^{(k)}(x_k+\beta_i), & i = k \\
\eta_{i-1}^{(k)}\alpha_{i-1} - \eta_i^{(k)}(x_k+\beta_i) + \eta_{i+1}^{(k)}\gamma_{i+1}, & 1 \le i < k \\
-\eta_0^{(k)}(x_k+\beta_0) + \eta_1^{(k)}\gamma_1 & i = 0
\end{cases}
$$
(3.30)

to compute the coefficients for $\pi_k(x)$, $k = 1 \ldots n$. Alongside this computation, we can also compute the coefficients of a sequence of polynomial

$g_k(x)$ of increasing degree k

$$g_k(x) = \sum_{i=0}^{k} c_i^{(k)} p_i(x)$$

initializing with $c_0^{(0)} = a_0$, where the a_i are still the Newton coefficients computed and used above. The subsequent coefficients $c_i^{(k)}$, $k = 1 \ldots n$ are computed using

$$c_i^{(k)} = \begin{cases} \eta_i^{(k)} a_k, & i = k \\ c_i^{(k-1)} + \eta_i^{(k)} a_k & 0 \le i < k. \end{cases} \tag{3.31}$$

This *incremental* construction of the coefficients, which is equivalent to effecting the summation of the weighted Newton polynomials in Equation 3.9 and referred to by Björck and Pereyra as the "progressive algorithm", can be used to efficiently update an interpolation. If the coefficients $\eta_i^{(n)}$ and $c_i^{(n)}$ are stored and a new node x_{n+1} and function value f_{n+1} are added to the data, a new coefficient a_{n+1} can be computed per Equation 3.10. The coefficients $\eta_i^{(n+1)}$ are then computed per Equation 3.30 and, finally, the $c_i^{(n)}$ are updated per Equation 3.31, resulting in the coefficients $c_i^{(n+1)}$ of the updated interpolation polynomial $g_{n+1}(x)$. We can re-write the recursion for the coefficients $\eta_i^{(k)}$ of the Newton polynomials in matrix-vector notation as

$$\left(\mathbf{T}^{(k+1)} - \underline{\mathbf{I}}_0 x_k \right) \boldsymbol{\eta}^{(k)} = \boldsymbol{\eta}^{(k+1)}$$

where $\mathbf{T}^{(k+1)}$ is the $(k+2) \times (k+1)$ tri-diagonal matrix

$$\mathbf{T}^{(k+1)} = \begin{pmatrix} -\beta_0 & \gamma_1 & & & \\ \alpha_0 & -\beta_1 & \gamma_2 & & \\ & \ddots & \ddots & \ddots & \\ & & \alpha_{k-2} & -\beta_{k-1} & \gamma_k \\ & & & \alpha_{k-1} & -\beta_k \\ & & & & \alpha_k \end{pmatrix}$$

and $\underline{\mathbf{I}}_0 x_k$ is a $(k+2) \times (k+1)$ matrix with x_k in the diagonal and zeros

---

Algorithm 4 Incremental construction of $g_n(x)$

1: $c_0^{(0)} \leftarrow f_0$ (*init* $\mathbf{c}^{(0)}$)

2: $\eta_0^{(1)} \leftarrow -x_0 - \beta_0$, $\eta_1^{(1)} \leftarrow \alpha_0$ (*init* $\boldsymbol{\eta}^{(1)}$)

3: **for** $k = 1 \ldots n$ **do**

4:     $v_0 \leftarrow 0$, $v_1 \leftarrow x_k$ (*init* $\mathbf{v}$)

5:     **for** $i = 2 \ldots k$ **do**

6:        $v_i \leftarrow \left((x_k + \beta_{i-1})v_{i-1} - \gamma_{i-1}v_{i-2} \right) / \alpha_{i-1}$ (*compute the* v_i)

7:     **end for**

8:     $g_k \leftarrow \mathbf{v}(0 : k-1)^\mathsf{T} \mathbf{c}^{(k-1)}$ (*compute* $g_{k-1}(x_k)$)

9:     $\pi_k \leftarrow \mathbf{v}^\mathsf{T} \boldsymbol{\eta}^{(k)}$ (*compute* $\pi_k(x_k)$)

10:     $a_k \leftarrow (f_k - g_k)/\pi_k$ (*compute* a_k, *Equation 3.32*)

11:     $\mathbf{c}^{(k)} \leftarrow [\mathbf{c}^{(k-1)}; 0] + a_k \boldsymbol{\eta}^{(k)}$ (*compute the new* $\mathbf{c}^{(k)}$, *Equation 3.31*)

12:     $\boldsymbol{\eta}^{(k+1)} \leftarrow [0; \underline{\alpha}(0 : k). * \boldsymbol{\eta}^{(k)}] - [(x_k + \underline{\beta}(0 : k)). * \boldsymbol{\eta}^{(k)}; 0] +$

         $[\underline{\gamma}(1 : k). * \boldsymbol{\eta}^{(k)}(1 : k); 0; 0]$

         (*compute the new* $\boldsymbol{\eta}^{(k+1)}$, *Equation 3.30*)

13: **end for**

---

elsewhere

$$
\underline{\mathbf{I}}_0 x_k = \begin{pmatrix} x_k & & & \\ & x_k & & \\ & & \ddots & \\ & & & x_k \\ 0 & 0 & \ldots & 0 \end{pmatrix}.
$$

The vectors $\boldsymbol{\eta}^{(k)} = (\eta_0^{(k)}, \eta_1^{(k)}, \ldots, \eta_k^{(k)})^\mathsf{T}$ and $\boldsymbol{\eta}^{(k+1)} = (\eta_0^{(k+1)}, \eta_1^{(k+1)}, \ldots, \eta_{k+1}^{(k+1)})^\mathsf{T}$ contain the coefficients of the kth and $(k+1)$st Newton polynomial.

Given the vector of coefficients $\mathbf{c}^{(n)} = (c_0^{(n)}, c_1^{(n)}, \ldots, c_n^{(n)})^\mathsf{T}$ of an interpolation polynomial $g_n(x)$ of degree n and the vector of coefficients $\boldsymbol{\eta}^{(n+1)}$ of the $(n+1)$st Newton polynomial over the $n+1$ nodes, we can *update* the interpolation, *i.e.* add a new node x_{n+1} and function value f_{n+1}, as follows: We first compute the new Newton interpolation coefficient a_k using the divided differences as in Equation 3.10 or, alternatively, by evaluating

$$
a_{n+1} = \frac{f_{n+1} - g_n(x_{n+1})}{\pi_{n+1}(x_{n+1})} \tag{3.32}
$$

using the coefficients and the three-term recurrence relation to evaluate both $g_n(x_{n+1})$ and $\pi_{n+1}(x_{n+1})$. The coefficient a_{n+1} is chosen such that the new interpolation constraint

$$g_{n+1}(x_{n+1}) = g_n(x_{n+1}) + a_{n+1}\pi_{n+1}(x_{n+1}) = f_{n+1}$$

is satisfied by construction. Note that since $\pi_{n+1}(x_i) = 0$ for $i = 0 \ldots n$, the addition of any multiple of $\pi_{n+1}(x)$ to $g_n(x)$ does not affect the interpolation at the other nodes at all.

This expression for a_{n+1} is used instead of the divided difference since we have not explicitly stored the previous a_i, $i = 0 \ldots n$, which are needed for the recursive computation of the latter.

We then update the coefficients of the interpolating polynomial

$$\mathbf{c}^{(n+1)} = \begin{pmatrix} \mathbf{c}^{(n)} \\ 0 \end{pmatrix} + a_{n+1}\boldsymbol{\eta}^{(n+1)} \tag{3.33}$$

and then the coefficients of the Newton polynomial

$$\boldsymbol{\eta}^{(n+2)} = \left(\mathbf{T}^{(n+2)} - \underline{\mathbf{I}}_0 x_{n+1}\right)\boldsymbol{\eta}^{(n+1)} \tag{3.34}$$

such that it is ready for further updates. Starting with $\eta_0^{(0)} = 1$ and $n = 0$, this update can be used to construct $g_n(x)$ by adding each x_i and f_i, $i = 0 \ldots n$, successively.

The complete algorithm is shown in Algorithm 4. The addition of each nth node requires $\mathcal{O}(n)$ operations, resulting in a total of $\mathcal{O}(n^2)$ operations for the construction of an n-node interpolation.

This is essentially the progressive algorithm of Björck and Pereyra, yet instead of storing the Newton coefficients a_i, we store the coefficients $\eta_i^{(n+1)}$ of the last Newton polynomial. This new representation offers no obvious advantage for the update, other than that it can be easily *reversed*:

Given an interpolation over a set of $n + 1$ nodes x_i and function values f_i, $i = 0 \ldots n$ defined by the coefficients $c_i^{(n)}$ and given the coefficients $\eta_i^{(n+1)}$ of the $(n + 1)st$ Newton polynomial over the same nodes, we can *downdate* the interpolation, *i.e.* remove the function value f_j at the node x_j. The resulting polynomial of degree $n - 1$ will still interpolate the remaining n nodes.

We begin by removing the root x_j from the $(n+1)$st Newton polynomial by solving

$$\left(\mathbf{T}^{(n+1)} - \mathbf{I}_0 x_j\right) \boldsymbol{\eta}^{(n)} = \boldsymbol{\eta}^{(n+1)}$$

for the vector of coefficients $\boldsymbol{\eta}^{(n)}$. If x_j is a root of $\pi_{n+1}(x)$, the system is over-determined yet has a unique solution[3]. We can therefore remove the first row of $(\mathbf{T}^{(n+1)} - \mathbf{I}_0 x_j)$ and the first entry of $\boldsymbol{\eta}^{(n+1)}$, resulting in the upper-tridiagonal system of linear equations

$$
\begin{pmatrix}
\alpha_0 & -(x_j+\beta_1) & \gamma_2 & & \\
& \ddots & \ddots & \ddots & \\
& & \alpha_{n-2} & -(x_j+\beta_{n-1}) & \gamma_n \\
& & & \alpha_{n-1} & -(x_j+\beta_n) \\
& & & & \alpha_n
\end{pmatrix}
\begin{pmatrix}
\eta_0^{(n)} \\
\eta_1^{(n)} \\
\vdots \\
\eta_n^{(n)}
\end{pmatrix}
=
\begin{pmatrix}
\eta_1^{(n+1)} \\
\eta_2^{(n+1)} \\
\vdots \\
\eta_{n+1}^{(n+1)}
\end{pmatrix}
$$
(3.35)

which can be conveniently solved in $\mathcal{O}(n)$ using back-substitution.

Once we have our downdated $\boldsymbol{\eta}^{(n)}$, and thus the downdated Newton polynomial $\pi_n(x)$, we can downdate the coefficients of $g_n(x)$ by computing

$$g_{n-1}(x) = g_n(x) - a_j^\star \pi_n(x)$$

where the Newton coefficient $a_j^\star$ would need to be re-computed from the divided difference over all nodes *except* x_j. We can avoid this computation by noting that $g_{n-1}(x)$ has to be of degree $n-1$ and therefore the highest coefficient of $g_n(x)$, $c_n^{(n)}$, must disappear. This is the case when

$$c_n^{(n-1)} = c_n^{(n)} - a_j^\star \eta_n^{(n)} = 0$$

and therefore

$$a_j^\star = \frac{c_n^{(n)}}{\eta_n^{(n)}}.$$

Using this $a_j^\star$, we can the compute the coefficients of $g_{n-1}(x)$ as

$$c_i^{(n-1)} = c_i^{(n)} - \frac{c_n^{(n)}}{\eta_n^{(n)}} \eta_i^{(n)}, \quad i = 1 \ldots n-1. \tag{3.36}$$

---

[3]Note that the $n \times (n+1)$ matrix $\left(\mathbf{T}^{(n+1)} - \mathbf{I}_0 x_j\right)^\mathsf{T}$ has rank n and the null space $\mathbf{p}(x_j) = (p_0(x_j), p_1(x_j), \ldots, p_{n+1}(x_j))^\mathsf{T}$ since for $\mathbf{v} = \left(\mathbf{T}^{(n+1)} - \mathbf{I}_0 x_j\right)^\mathsf{T} \mathbf{p}(x_j)$, $v_i = \alpha_i p_{i+1}(x_j) - (x_j + \beta_i) p_i(x_j) + \gamma_i p_{i-1}(x_j) = 0$ by the definition in Equation 3.14 and the right-hand side $\boldsymbol{\eta}^{(n+1)}$ is consistent.

The whole process is shown in Algorithm 5. The downdate of an n-node interpolation requires $\mathcal{O}(n)$ operations.

Algorithm 5 Remove a function value f_j at the node x_j from the interpolation given by the coefficients $\mathbf{c}^{(n)}$

1: $\eta_n^{(n)} \leftarrow \eta_{n+1}^{(n+1)}/\alpha_n$ (*compute $\boldsymbol{\eta}^{(n)}$ from $\boldsymbol{\eta}^{(n+1)}$*)

2: $\eta_{n-1}^{(n)} \leftarrow \left(\eta_n^{(n+1)} + (x_j + \beta_n)\eta_n^{(n)}\right)/\alpha_{n-1}$

3: **for** $i = n - 2 \ldots 0$ **do**

4: $\quad \eta_i^{(n)} \leftarrow \left(\eta_{i+1}^{(n+1)} + (x_j + \beta_{i+1})\eta_{i+1}^{(n)} - \gamma_{i+2}\eta_{i+2}^{(n)}\right)/\alpha_i$

5: **end for**

6: $a_j \leftarrow c_n^{(n)}/\eta_n^{(n)}$ (*compute the coefficient a_j*)

7: $\mathbf{c}^{(n-1)} \leftarrow \mathbf{c}^{(n)} - a_j\boldsymbol{\eta}^{(n)}$ (*compute the new coefficients $\mathbf{c}^{(n-1)}$*)

3.2.2 A New Algorithm for the Construction of Interpolations

Returning to the representation in Equation 3.17, we can try to solve the Vandermonde-like system of linear equations directly. The matrix has some special characteristics which can be exploited to achieve better performance and stability than when using Gaussian elimination or even the algorithms of Björck and Pereyra, Higham or the one described in the previous section.

We start by de-composing the $(n+1) \times (n+1)$ Vandermonde-like matrix $\mathbf{P}^{(n)}$ as follows:

$$
\mathbf{P}^{(n)} = \left(
\begin{array}{c|c}
\mathbf{P}^{(n-1)} & \mathbf{p}^{(n)} \\
\hline
\mathbf{q}^{\mathsf{T}} & p_n(x_n)
\end{array}
\right).
$$

The sub-matrix $\mathbf{P}^{(n-1)}$ is a Vandermonde-like matrix analogous to $\mathbf{P}^{(n)}$. The column $\mathbf{p}^{(n)}$ contains the nth polynomial evaluated at the nodes x_i,

$i = 0 \ldots n - 1$

$$\mathbf{p}^{(n)} = \begin{pmatrix} p_n(x_0) \\ p_n(x_1) \\ \vdots \\ p_n(x_{n-1}) \end{pmatrix}$$

and the vector $\mathbf{q}^{\mathsf{T}}$ contains the first $0 \ldots n - 1$ polynomials evaluated at the node x_n

$$\mathbf{q}^{\mathsf{T}} = (p_0(x_n), p_1(x_n), \ldots, p_{n-1}(x_n)) .$$

Inserting this into the product in Equation 3.18, we obtain

$$\left(\begin{array}{c|c} \mathbf{P}^{(n-1)} & \mathbf{p}^{(n)} \\ \hline \mathbf{q}^{\mathsf{T}} & p_n(x_n) \end{array} \right) \left(\begin{array}{c} \mathbf{c}^{(n-1)} \\ \hline c_n \end{array} \right) = \left(\begin{array}{c} \mathbf{f}^{(n-1)} \\ \hline f_n \end{array} \right)$$

which, when applied, results in the equations

$$\begin{aligned} \mathbf{P}^{(n-1)} \mathbf{c}^{(n-1)} + \mathbf{p}^{(n)} c_n &= \mathbf{f}^{(n-1)} \\ \mathbf{q}^{\mathsf{T}} \mathbf{c}^{(n-1)} + p_n(x_n) c_n &= f_n. \end{aligned} \tag{3.37}$$

where the vectors $\mathbf{c}^{(n-1)} = (c_0, c_1, \ldots, c_{n-1})^{\mathsf{T}}$ and $\mathbf{f}^{(n-1)} = (f_0, f_1, \ldots, f_{n-1})^{\mathsf{T}}$ contain the first n coefficients or function values respectively.

Before solving Equation 3.37, we note that the matrix $\mathbf{P}^{(n-1)}$ contains the first $0 \ldots n-1$ polynomials evaluated at the same n nodes each. Similarly, $\mathbf{q}^{\mathsf{T}}$ contains the same polynomials evaluated at the node x_n. Since the polynomials in the columns are of degree $< n$ and they are evaluated at n points, $\mathbf{P}^{(n-1)}$ actually contains enough data to extrapolate the values of these polynomials at x_n. Using the Lagrange interpolation described in Equation 3.5, we can write

$$\begin{aligned} p_i(x_n) &= \sum_{j=0}^{n-1} \ell_j^{(n)}(x_n) p_i(x_j) \\ q_i &= \sum_{j=0}^{n-1} \ell_j^{(n)}(x_n) P_{j,i}^{(n-1)} \end{aligned} \tag{3.38}$$

where the $\ell_j^{(n)}(x)$ are the Lagrange polynomials over the first n nodes x_i, $i = 0 \ldots n - 1$. We write Equation 3.38 as

$$\mathbf{q}^{\mathsf{T}} = \underline{\ell}^{(n)} \mathbf{P}^{(n-1)} \tag{3.39}$$

where the entries of the $1 \times n$ vector $\underline{\ell}^{(n)}$ are

$$\ell_i^{(n)} = \ell_i(x_n).$$

The entries of $\underline{\ell}^{(n)}$ can be computed recursively. Using the definition in Equation 3.6, we can write the individual $\ell_i^{(n)}$ as

$$\ell_i^{(n)} = \prod_{\substack{j=0 \\ j \neq i}}^{n-1} \frac{x_n - x_j}{x_i - x_j}.$$

If we define

$$w_n = \prod_{j=0}^{n-1} (x_n - x_j)$$

we can re-write $\ell_i^{(n)}$ as

$$\ell_i^{(n)} = \frac{w_n}{x_n - x_i} \left[\prod_{\substack{j=0 \\ j \neq i}}^{n-1} (x_i - x_j) \right]^{-1}. \tag{3.40}$$

We can extract the product on the right-hand side using the previous $\ell_i^{(n-1)}$ and w_{n-1} over the first $0 \ldots n - 2$ nodes:

$$\ell_i^{(n-1)} = \frac{w_{n-1}}{x_{n-1} - x_i} \left[\prod_{\substack{j=0 \\ j \neq i}}^{n-2} (x_i - x_j) \right]^{-1}$$

$$\left[\prod_{\substack{j=0 \\ j \neq i}}^{n-2} (x_i - x_j) \right]^{-1} = \frac{\ell_i^{(n-1)}}{w_{n-1}} (x_{n-1} - x_i)$$

which, re-inserted into Equation 3.40, gives

$$\ell_i^{(n)} = \frac{w_n}{x_n - x_i} \frac{\ell_i^{(n-1)}}{w_{n-1}} \frac{x_{n-1} - x_i}{x_i - x_{n-1}}$$

$$= -\frac{\ell_i^{(n-1)}}{x_n - x_i} \frac{w_n}{w_{n-1}} \qquad (3.41)$$

for all $i < n - 1$. We then compute the last entry $i = n - 1$ using

$$\ell_{n-1}^{(n)} = \frac{w_n}{w_{n-1}(x_n - x_{n-1})}. \qquad (3.42)$$

Therefore, starting with $\ell_0^{(1)} = 1$, we can construct all the $\underline{\ell}^{(k)}$, $k = 2 \ldots n$ successively. Since, given $\underline{\ell}^{(k)}$, $k = 1 \ldots n$ the construction of each additional $\underline{\ell}^{(n+1)}$ requires $\mathcal{O}(n)$ operations, the construction of all the $\underline{\ell}^{(k)}$, $k = 1 \ldots n$ requires a total of $\mathcal{O}(n^2)$ operations.

Returning to the Vandermonde-like matrix, inserting Equation 3.39 into Equation 3.37, we obtain

$$\mathbf{P}^{(n-1)}\mathbf{c}^{(n-1)} + \mathbf{p}^{(n)}c_n = \mathbf{f}^{(n-1)}$$

$$\underline{\ell}^{(n)}\mathbf{P}^{(n-1)}\mathbf{c}^{(n-1)} + p_n(x_n)c_n = f_n. \qquad (3.43)$$

Multiplying the first line in Equation 3.43 with $\underline{\ell}^{(n)}$ from the left we obtain

$$\underline{\ell}^{(n)}\mathbf{P}^{(n-1)}\mathbf{c}^{(n-1)} + \underline{\ell}^{(n)}\mathbf{p}^{(n)}c_n = \underline{\ell}^{(n)}\mathbf{f}^{(n-1)}$$

$$\underline{\ell}^{(n)}\mathbf{P}^{(n-1)}\mathbf{c}^{(n-1)} + p_n(x_n)c_n = f_n. \qquad (3.44)$$

and subtracting the bottom equation from the top one we obtain

$$\underline{\ell}^{(n)}\mathbf{p}^{(n)}c_n - p_n(x_n)c_n = \underline{\ell}^{(n)}\mathbf{f}^{(n-1)} - f_n$$

from which we can finally isolate the coefficient c_n:

$$c_n = \frac{\underline{\ell}^{(n)}\mathbf{f}^{(n-1)} - f_n}{\underline{\ell}^{(n)}\mathbf{p}^{(n)} - p_n(x_n)}. \qquad (3.45)$$

Having computed c_n, we can now re-insert it into the first Equation in Equation 3.43, resulting in the new system

$$\mathbf{P}^{(n-1)}\mathbf{c}^{(n-1)} = \mathbf{f}^{(n-1)} - \mathbf{p}^{(n)}c_n \qquad (3.46)$$

in the remaining coefficients $\mathbf{c}^{(n-1)}$.

Applying this computation recursively to $\mathbf{P}^{(n)}$, $\mathbf{P}^{(n-1)}$, ..., $\mathbf{P}^{(1)}$ we can compute the interpolation coefficients $\mathbf{c}^{(n)}$. The final coefficient c_0 can be computed as

$$c_0 = f_0/P_{1,1}^{(0)}.$$

The complete algorithm is shown in Algorithm 6. Since the construction of the $\underline{\ell}^{(k)}$, $k = 1 \ldots n$ requires $\mathcal{O}(n^2)$ operations (for each $\underline{\ell}^{(k)}$, $\mathcal{O}(k)$ operations are required to compute w_k and $\mathcal{O}(k)$ are required to compute the new entries) and the evaluation of Equation 3.45 requires $\mathcal{O}(k)$ operations for each c_k, $k = n \ldots 0$, the total cost of the algorithm is in $\mathcal{O}(n^2)$ operations.

Algorithm 6 Direct construction of $g_n(x)$

1: $\mathbf{p}^{(0)} \leftarrow 0$, $\mathbf{p}^{(1)} \leftarrow \mathbf{x}$ (*init* $\mathbf{P}^{(n)}$)
2: **for** $i = 2 \ldots n$ **do**
3:     $\mathbf{p}^{(i)} \leftarrow \left((\mathbf{x} + \beta_{i-1}) . * \mathbf{p}^{(i-1)} - \gamma_{i-1}\mathbf{p}^{(i-2)} \right)/\alpha_{i-1}$ (*fill* $\mathbf{P}^{(n)}$)
4: **end for**
5: $\ell_0^{(1)} \leftarrow 1$, $w_1 = x_1 - x_0$ (*init* $\ell_0^{(1)}$ *and* w_1)
6: **for** $i = 2 \ldots n$ **do**
7:     $w_i \leftarrow 1$ (*construct* w_i)
8:     **for** $j = 0 \ldots i - 1$ **do**
9:         $w_i \leftarrow w_i(x_i - x_j)$
10:     **end for**
11:     **for** $j = 0 \ldots i - 2$ **do**
12:         $\ell_j^{(i)} \leftarrow -\dfrac{\ell_j^{(i-1)}}{x_i - x_j}\dfrac{w_i}{w_{i-1}}$ (*compute the* $\ell_j^{(i)}$, *Equation 3.41*)
13:     **end for**
14:     $\ell_{i-1}^{(i)} \leftarrow \dfrac{w_i}{w_{i-1}(x_i - x_{i-1})}$ (*compute* $\ell_{i-1}^{(i)}$, *Equation 3.42*)
15: **end for**
16: **for** $i = n \ldots 1$ **do**
17:     $c_i \leftarrow \dfrac{\ell^{(i)}\mathbf{f}(0:i-1) - f_i}{\ell^{(i)}\mathbf{p}^{(i-1)}(0:i-1) - p_i(x_i)}$ (*compute coefficient* c_i, *Equation 3.45*)
18:     $\mathbf{f} \leftarrow \mathbf{f} - c_i\mathbf{p}^{(i)}$ (*update the right-hand side* $\mathbf{f}$)
19: **end for**
20: $c_0 \leftarrow f_0/\mathbf{p}^{(0)}(0)$ (*compute the final* c_0)

Note that, as opposed to the incremental Algorithm 4 presented in Section 2.1, this algorithm can not be extended to update or downdate an interpolation. It has an advantage, however, when multiple right-hand sides, *i.e.* interpolations over the same set of nodes, are to be computed. In such a case, the vectors $\underline{\ell}^{(k)}$, $k = 1 \dots n$ only need to be computed once (Lines 5 to 15 of Algorithm 6). For any new vector $\mathbf{f}$, only the Lines 16 to 20 need to be re-evaluated.

3.2.3   Numerical Tests

To assess the stability of the two new interpolation routines described herein, we will follow the methodology used by Higham in [46]. Higham defines a set of interpolations consisting of all combinations of the nodes

$$
\begin{array}{lll}
\text{A1:} & x_i = -\cos(i\pi/n), & (\text{extrema of } T_n(x)) \\
\text{A2:} & x_i = -\cos\left[(i + \tfrac{1}{2})\pi/(n+1)\right], & (\text{zeros of } T_{n+1}(x)) \\
\text{A3:} & x_i = -1 + 2i/n, & (\text{equidistant on } [-1, 1]) \\
\text{A4:} & x_i = i/n, & (\text{equidistant on } [0, 1])
\end{array}
$$

with the right-hand sides

$$
\begin{array}{lll}
\text{F1:} & f_i = (-1)^i, \\
\text{F2:} & f = [1, 0, \dots, 0]^\mathsf{T} \\
\text{F3:} & f_i = 1/(1 + 25x_i^2)
\end{array}
$$

for $i = 0 \dots n$.

To avoid instabilities caused by unfortunate orderings of the nodes x_i, the nodes and corresponding function values were re-ordered according to the same permutation that would be produced by Gaussian elimination with partial pivoting applied to the Vandermonde-like matrix as described by Higham in [47]. This ordering is optimal for the Björck-Pereyra and Higham algorithms and produces good results for the new algorithms described in the previous two sections.

For each combination of nodes and right-hand sides, we compute, following Higham, the coefficients $\mathbf{c}$ for the Chebyshev base (see Equation 3.16) using for $n = 5$, 10, 20 and 30 and compute the quantities

$$
\text{ERR} = \frac{\|\mathbf{c} - \mathbf{c}^\star\|_2}{u\|\mathbf{c}^\star\|_2}, \quad \text{RES} = \frac{\|\mathbf{f} - \mathbf{Pc}\|_2}{u\|\mathbf{c}^\star\|_2},
$$

	GE		BP/H		INCR		DIRECT		DEL	
n	ERR	RES	ERR	RES	ERR	RES	ERR	RES	ERR	RES
5	0.00	0.50	2.00	3.94	**3.20**	**4.30**	0.00	0.50	0.00	0.72
10	3.20	5.32	**1.20e1**	**2.79e1**	7.76	2.21e1	2.26	7.00	0.00	2.47
20	7.28	2.16e1	**1.61e2**	**5.27e2**	8.92	3.64e1	9.14	2.60e1	0.00	3.40e1
30	2.61	1.03e1	**6.72e2**	**2.65e3**	2.08e1	1.11e2	3.51	1.60e1	0.00	8.76e1

Table 3.1: Results for problem A1/F1.

where c^* is the exact solution and u is the unit roundoff [4] as defined in [41, Section 2.4.2].

Results were computed using Gaussian elimination (GE[5]), Higham's extension of the algorithm of Björck and Pereyra (BP/H[6]), the incremental Algorithm 4 (INCR) and the direct Algorithm 6 (DIRECT). The exact values were computed in Maple [9] with 50 decimal digits of precision using the interp function therein.

Results were also computed for the interpolation downdate (DEL) described in Algorithm 5. Starting from c^* and η^*, the exact coefficients for the interpolation $g_n(x)$ and the Newton polynomial $\pi_{n+1}(x)$ respectively, we compute the coefficients $c^{(n-1)}$ and $\eta^{(n)}$ for $g_{n-1}(x)$ and $\pi_n(x)$, resulting from the removal of the rightmost function value f_k at x_k, $k = \arg\max_i x_i$. The exact coefficients $\hat{c}^*$ *after* deletion were computed and used to compute the quantities ERR and RES.

The results are shown in Tables 3.1 to 3.12. For each n, the largest values for ERR and RES are highlighted. For the problem sets over the nodes A1 and A2 (Tables 3.1 to 3.6), the condition of the Vandermonde-like matrix is always ≤ 2 [37], resulting in very small errors for Gaussian elimination. The Björck-Pereyra/Higham generates larger residuals than both the incremental and direct algorithms for both sets of nodes. The values for ERR, however, are usually within the same order of magnitude for the three algorithms.

---

[4] All results were computed using IEEE 754 double-precision arithmetic and hence $u \approx 2.2 \times 10^{-16}$.

[5] For the tests in this section, Matlab's backslash-operator, which uses partial pivoting, was used. In cases where the matrix is rank-deficient, a minimum-norm solution is returned.

[6] Algorithm 1 in [46] was implemented in Matlab.

n	GE ERR	RES	BP/H ERR	RES	INCR ERR	RES	DIRECT ERR	RES	DEL ERR	RES
5	0.97	2.40	**1.82**	**3.19**	0.93	2.07	0.88	1.68	0.00	1.70
10	2.22	4.19	**1.37e1**	**3.77e1**	4.94	1.32e1	1.93	3.00	0.00	3.80
20	2.11e1	8.96	**9.93e1**	**3.47e2**	2.24e1	4.80e1	1.80e1	1.27e1	0.00	5.39e1
30	3.63e1	1.36e1	**1.27e2**	**4.84e2**	5.55e1	2.27e2	4.31e1	2.88e1	0.00	1.19e2

Table 3.2: Results for problem A1/F2.

n	GE ERR	RES	BP/H ERR	RES	INCR ERR	RES	DIRECT ERR	RES	DEL ERR	RES
5	1.26	2.67	1.16	2.27	1.26	2.70	**1.43**	**2.99**	0.00	1.18
10	2.12	5.30	**7.27**	**1.66e1**	1.83	3.63	1.19	1.87	0.32	2.67
20	1.13	4.81	**8.28**	**2.68e1**	1.78	6.74	3.02	1.01e1	0.05	5.25
30	1.99	7.35	**6.33**	**2.53e1**	1.14	5.86	2.55	1.04e1	0.61	6.09

Table 3.3: Results for problem A1/F3.

n	GE ERR	RES	BP/H ERR	RES	INCR ERR	RES	DIRECT ERR	RES	DEL ERR	RES
5	3.55	0.70	**5.07**	**7.31**	3.79	1.58	4.96	4.96	1.48	3.16
10	1.19e1	3.65	**2.03e1**	**3.31e1**	8.34	1.08e1	1.16e1	2.24	7.23	7.47
20	1.66e1	1.23e1	4.64e1	**1.63e2**	**5.57e1**	1.41e1	1.61e1	2.58	1.69e1	2.41e1
30	6.48e1	2.28e1	**1.24e2**	**4.36e2**	4.45e1	2.29e2	6.49e1	5.98	4.71e1	9.82e1

Table 3.4: Results for problem A2/F1.

n	GE ERR	RES	BP/H ERR	RES	INCR ERR	RES	DIRECT ERR	RES	DEL ERR	RES
5	2.57	3.29	**2.88**	**4.52**	2.35	0.97	2.30	1.20	0.60	3.25
10	5.27	3.89	**1.67e1**	**4.59e1**	4.94	5.23	3.94	3.75	4.65	1.03e1
20	8.40	8.80	**5.09e1**	**1.63e2**	3.89e1	9.85e1	8.44	4.04	1.15e1	2.43e1
30	3.39e1	1.95e1	**1.17e2**	**4.31e2**	3.00e1	1.99e1	3.42e1	2.55e1	3.27e1	1.13e2

Table 3.5: Results for problem A2/F2.

n	GE ERR	RES	BP/H ERR	RES	INCR ERR	RES	DIRECT ERR	RES	DEL ERR	RES
5	**1.44**	0.11	1.40	**1.71**	1.17	1.45	1.12	1.67	0.00	2.76
10	2.73	3.22	**6.06**	**1.14e1**	2.86	3.75	3.64	6.34	6.56	7.47
20	1.52	5.59	**7.34**	**2.42e1**	2.06	6.34	3.92	1.34e1	1.27e1	1.81e1
30	2.81	1.19e1	**7.24**	**3.08e1**	1.65	6.07	2.79	9.46	1.41e1	2.94e1

Table 3.6: Results for problem A2/F3.

n	GE ERR	RES	BP/H ERR	RES	INCR ERR	RES	DIRECT ERR	RES	DEL ERR	RES
5	1.41	0.58	**6.12**	**1.01e1**	2.04	0.97	1.48	0.73	0.60	0.78
10	2.99	2.50	**1.18e1**	**2.00e1**	2.16	1.52	2.62	3.11	0.50	1.42
20	**2.26e3**	4.85	2.01e1	**7.19e1**	3.58e1	1.02e1	6.23e1	3.97	0.55	2.77
30	**1.39e6**	9.51	3.90e1	**1.34e2**	5.42e4	2.60e1	3.07e2	7.33	0.55	2.20

Table 3.7: Results for problem A3/F1.

n	GE ERR	RES	BP/H ERR	RES	INCR ERR	RES	DIRECT ERR	RES	DEL ERR	RES
5	1.02	1.98	**2.19**	**3.53**	0.69	0.80	0.73	1.40	0.40	0.86
10	**4.91**	7.58	3.83	**1.03e1**	1.05	1.82	1.61	2.46	0.47	0.85
20	**3.65e3**	1.07e1	8.32	**2.54e1**	5.73e2	2.94	1.29	4.25	0.63	2.26
30	**4.02e5**	9.93	3.23e1	**1.09e2**	1.35e5	1.25e1	4.98	7.32	0.65	3.48

Table 3.8: Results for problem A3/F2.

n	GE ERR	RES	BP/H ERR	RES	INCR ERR	RES	DIRECT ERR	RES	DEL ERR	RES
5	1.48	0.86	**1.87**	**1.98**	1.78	1.89	1.38	0.36	0.00	2.23
10	2.31	3.28	**8.96**	**2.76e1**	2.00	2.97	3.27	3.80	0.45	1.12
20	**2.30e3**	5.53	3.81e1	**7.38e1**	1.12e2	1.24e1	3.13e1	5.11	0.47	3.39
30	**1.39e6**	1.10e1	2.28e2	**1.88e2**	5.29e4	2.69e1	2.41e2	4.78	0.49	2.29

Table 3.9: Results for problem A3/F3.

n	GE ERR	GE RES	BP/H ERR	BP/H RES	INCR ERR	INCR RES	DIRECT ERR	DIRECT RES	DEL ERR	DEL RES
5	2.01e2	0.55	0.55	**0.76**	**5.40e2**	0.74	1.57e1	0.48	3.06	3.80
10	2.49e4	0.84	0.45	0.76	**3.56e6**	**1.67**	8.20e2	0.60	1.94	3.55
20	**2.77e15**	**2.53**	1.87	1.69	5.08e14	1.78	6.00e7	1.63	5.93	1.41e1
30	4.50e15	0.00	4.44	1.16	−	−	3.47e11	1.60	8.93	3.82e1

Table 3.10: Results for problem A4/F1.

n	GE ERR	GE RES	BP/H ERR	BP/H RES	INCR ERR	INCR RES	DIRECT ERR	DIRECT RES	DEL ERR	DEL RES
5	1.82e2	0.48	0.55	**0.65**	**3.64e2**	0.37	1.12e1	0.45	1.34	2.27
10	1.35e5	0.63	0.40	0.85	**3.77e6**	**1.55**	6.28e2	0.75	0.99	2.01
20	**2.81e15**	**3.64**	0.71	1.20	6.57e14	0.93	4.05e7	1.57	1.83	3.30
30	4.50e15	0.00	0.40	1.27	−	−	1.37e11	1.73	2.88	1.27e1

Table 3.11: Results for problem A4/F2.

n	GE ERR	GE RES	BP/H ERR	BP/H RES	INCR ERR	INCR RES	DIRECT ERR	DIRECT RES	DEL ERR	DEL RES
5	2.41e2	0.23	3.96e1	0.72	**8.24e2**	**0.73**	8.23	0.62	9.95	1.13e1
10	3.79e5	**1.38**	1.55e3	0.88	**4.94e6**	1.35	3.04e2	0.55	0.26	0.28
20	**2.81e15**	**3.74**	4.84e6	1.21	9.48e14	0.87	3.56e7	1.03	1.45	2.58
30	4.50e15	0.00	1.02e11	1.47	−	−	1.18e11	1.49	4.56	1.97e1

Table 3.12: Results for problem A4/F3.

For the nodes A3 (Tables 3.7 to 3.9), the condition number of the Vandermonde-like matrix is 5.11e6 for $n = 30$, resulting in the errors of approximately that magnitude when Gaussian elimination is used. In general, both the Björck-Pereyra/Higham and the direct algorithm generate
smaller errors and residues than Gaussian elimination. The errors for the
incremental algorithm are due to cancellation while evaluating $g_n(x_{n+1})$
for Equation 3.32, since the intermediate coefficients $\mathbf{c}^{(k)}$ are several orders of magnitude larger than the result[7].

Finally, the condition number of the Vandermonde-like matrix for the
nodes A4 is 4.26e16 for $n = 30$, making it numerically singular and
thus resulting in the complete failure of Gaussian elimination. Note that
since in such cases Matlab's backslash-operator computes the minimum
norm solution, the resulting residual error RES is quite small. For the
first two right-hand sides F1 and F2, the Björck-Pereyra/Higham algorithm performs significantly better than the two new algorithms, since
the magnitude of the intermediate coefficients does not vary significantly.
For the right-hand side F3, however, the errors are larger, caused by
truncation in computing the Newton coefficients a_i. The incremental algorithm fails completely for all right-hand sides since the intermediate
and final coefficients $\mathbf{c}^{(k)}$, $k \leq n$ are more than ten orders of magnitude
larger than the function values[8] and the numerical condition of $g_n(x_{n+1})$
in Equation 3.32 exceeds machine precision, resulting in numerical overflow. These relatively large coefficients also cause problems for the direct
algorithm when evaluating the right-hand side of Equation 3.46, where
the original function values are clobbered by the subtraction of the much
larger $\mathbf{p}^{(n)}c_n$.

The errors and residuals for the downdate algorithm are shown in the
rightmost columns of Tables 3.1 to 3.12. In general the errors and residues
of the downdate are relatively small for all test cases. The larger values,
e.g. for A2/F2, are due to cancellation in the final subtraction in Algorithm 5, Line 7.

---

[7]In [46], Higham shows that the coefficients can be written as the weighted sum of
any of the intermediate coefficients $c_i^{(n)} = \sum_j \mu_j c_j^{(k)}$, where the μ_j depend only on
the nodes and the coefficients of the three-term recurrence relation. If the μ_j are $\mathcal{O}(1)$
and the intermediate $c_j^{(k)}$ are much larger than the $c_i^{(n)}$, then cancellation is likely to
occur in the above sum.
[8]$\|\mathbf{c}^\star\| = 2.23e13$ for F3 and $n = 30$.

Chapter 4

Error Estimation

In this chapter we will start by discussing the different types of error estimators used in the adaptive quadrature algorithms discussed in Chapter 2. We will divide these error estimators using the explicit representation of the integrand described in Chapter 3, in two distinct categories and discuss their common features and shortcomings (Section 4.1).

We then derive two new error estimators based on using the L_2-norms of the interpolating polynomials on which each quadrature rule is based (Section 4.2).

Finally, the most prominent error estimators from Chapter 2, as well as the two new estimates presented here are tested against a set of problematic integrands (Section 4.3) and the results of these tests are analyzed (Section 4.4).

4.1 Categories of Error Estimators

In the following, we will distinguish between *linear* and *non-linear* error estimates.

4.1.1   Linear Error Estimators

We can group the different linear error estimators from Chapter 2 in the following categories:

1. $\varepsilon \sim \left| Q_n^{(m_1)}[a,b] - Q_n^{(m_2)}[a,b] \right|$: Error estimators based on the difference between two estimates of the same degree yet of different multiplicity (Kuncir [55], McKeeman [68, 71, 69], Lyness [61, 62], Malcolm and Simpson [66] and Forsythe *et al.* [29]).

2. $\varepsilon \sim |Q_{n_1}[a,b] - Q_{n_2}[a,b]|$: Error estimators based on the difference between two estimates of different degree (Patterson [80], Piessens *et al.* [82, 83], Hasegawa *et al.* [42], Berntsen and Espelid [4], Favati *et al.* [26] and Gander and Gautschi [34]).

3. $\varepsilon \sim \left| f^{(n)}(\xi) \right|$: Error estimators based on directly approximating the derivative in the analytic error term (Gallaher [30], Garribba *et al.* [35] and Ninomiya [74]).

4. $\varepsilon \sim |\tilde{c}_n|$: Error estimators based on the estimate of the highest-degree coefficient of the function relative to some orthogonal base (O'Hara and Smith [76, 77], Oliver [79], Berntsen and Espelid [5, 19, 20, 22, 23, 24]).

Already in 1985, Laurie [57] shows that the first three categories are, in essence, *identical*. Consider Kuncir's (see Section 2.2) error estimate from the **first** category,

$$\left| \mathsf{S}^{(1)}[a,b] - \mathsf{S}^{(2)}[a,b] \right|,$$

which can be viewed as a 5-point "rule" (or linear functional) over the nodes used by $\mathsf{S}^{(1)}[a,b]$ and $\mathsf{S}^{(2)}[a,b]$ (see Fig. 2.2).

Since both approximations evaluate polynomials of up to degree 3 exactly, their difference will be, when applied to polynomials of up to degree 3, zero. When applied to a polynomial of degree 4 or higher, the estimates will, in all but pathological cases, differ. This is, up to a constant factor, *exactly* what the *4th divided difference* over the same 5 nodes computes:

It evaluates polynomials of up to degree 3 to zero and polynomials of degree 4 and higher in all but pathological cases to some non-zero value[1]. The same can be said of error estimates from the **second** category, such as the one used by Piessens (see Section 2.12)

$$|\mathsf{G}_n[a,b] - \mathsf{K}_{2n+1}[a,b]|$$

where the Gauss quadrature rule $\mathsf{G}_n[a,b]$ integrates all polynomials of degree up to $2n-1$ exactly and its Kronrod extension $\mathsf{K}_{2n+1}[a,b]$ integrates all polynomials of degree up to $3n+1$ exactly. Since the approximation computed by these rules differ only for polynomials of degree $2n$ and higher, the combined "rule" over the $2n+1$ points behaves just as the $2n$th *divided difference* would.

This same principle also applies to Berntsen and Espelid's (see Section 2.18) improved error estimate

$$|\mathsf{G}_{2n+1}[a,b] - Q_{2n-1}[a,b]|$$

where the interpolatory quadrature rule $Q_{2n-1}[a,b]$ over $2n$ points integrates all polynomials up to degree $2n-1$ exactly, making the difference to $\mathsf{G}_{2n+1}[a,b]$ zero for all polynomials up to degree $2n-1$ and non-zero for all non-pathological polynomials of degree $2n$ or higher, which is equivalent to the $2n$th *divided difference* over those same points.

In all these cases, the divided differences are *unique* (*i.e.* the nth difference over $n+1$ points), as are the quadrature rules. They therefore *differ only by a constant factor*. As a consequence, the first and second categories, in which the difference between two quadrature rules is used to approximate the error, are both equivalent to the **third** category, in which the lowest degree derivative of the error expansion are approximated explicitly.

Not all error estimators in these categories, though, are identical up to a constant factor to the highest-degree divided differences over the same points. McKeeman's error estimator (see Section 2.3), for instance, approximates a 4th divided difference over 7 points, which is neither unique nor of the highest-possible degree. The same can be said of Forsythe,

---

[1]Note that this is also, up to a constant factor, the definition of a null-rule, as used by Berntsen and Espelid (see Section 2.19). Lyness [60], who originally introduced the concept of null-rules, creates them explicitly from the difference of two quadrature rules, as is done in these error estimates implicitly.

Malcolm and Moler's QUANC8 (see Section 2.7). Patterson's successive
Kronrod extensions (see Section 2.12) are also an exception since the
error estimates of the type

$$\left| \mathsf{K}_n[a, b] - \mathsf{K}_{2n+1}[a, b] \right|,$$

where $\mathsf{K}_n[a, b]$ integrates polynomials up to degree $(3n + 1)/2$ exactly,
compute only the $(3n - 1)/2$th difference over $2n + 1$ points, which is no
longer unique. In general, though, we can say that in all these methods
the approximated divided difference is of one degree higher than the
degree of the lowest-degree rule in the pair used to compute it.

In the **fourth** and final category we again find finite differences, namely
in Berntsen and Espelid's null rules (see Section 2.19), in which the
coefficients e_k relative to an orthogonal base are computed (see Equation 2.103). The highest-degree coefficient e_{n-1}, computed with the
$(n - 1)$st null rule over n nodes is, as Berntsen and Espelid themselves
note in [5], identical up to a constant factor to the $(n - 1)$st divided difference over the same nodes. This value is combined with the $(n - 2)$nd
divided difference (see Equation 2.105), itself identical only up to a linear
factor, to avoid phase effects, and used as an error estimate.

The same goes for the coefficients relative to *any* base computed over n
points, such as the coefficients $\tilde{c}_i$ of the Chebyshev polynomials used by
O'Hara and Smith (see Section 2.8) and Oliver (see Section 2.11). The
"rule" used to compute the highest-degree coefficients (Equation 2.24
and Equation 2.28) is identical up to a constant factor to the nth divided difference over the $n + 1$ nodes used. While O'Hara and Smith use
the highest-degree coefficient directly, Oliver uses $K^3 |\tilde{c}_{n-4}|$ (see Equa-
tion 2.54 and Equation 2.55), which is related (*i.e.* no longer identical up
to a constant factor) to the $(n - 4)$th divided difference.

We therefore establish that *all* linear error estimators presented in this
section are equivalent in that they all use one or more divided difference
approximations of the higher derivatives of the integrand. The quality of
the error estimate therefore depends on the quality of these approximations.

Using the difference of two interpolatory quadrature rules to compute
the divided differences, as is done in the first two categories of error
estimators, it is easy to show how problems can arise: Consider the case
of the smooth function in Fig. 4.1, the integral of which is approximated

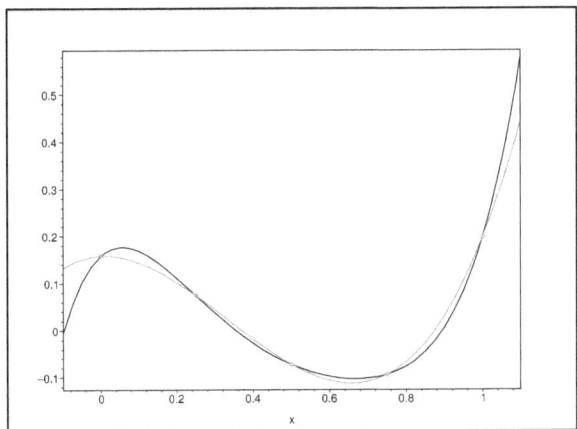

Figure 4.1: A smooth function (black) interpolated by $\tilde{f}_1(x)$ (green) using the nodes of $S^{(1)}[a,b]$ and $\tilde{f}_2(x)$ (red) using the nodes of $S^{(2)}[a,b]$.

over $[0,1]$ using both $S^{(1)}[a,b]$ and $S^{(2)}[a,b]$. Let $\tilde{f}_1(x)$ be the polynomial interpolated over the nodes of $S^{(1)}[a,b]$ and $\tilde{f}_2(x)$ that interpolated over the nodes of $S^{(2)}[a,b]$.

Although the interpolations $\tilde{f}_1(x)$ and $\tilde{f}_2(x)$ differ significantly, which is or should be a good indication that the degree of our quadrature rule is not sufficient, the integral of both $\tilde{f}_1(x)$ and $\tilde{f}_2(x)$ in $[0,1]$ is identical and the 4th divided difference over the nodes used, and thus the error estimate, gives 0.

In smooth functions, this type of problem may occur when the integrand is not "sufficiently smooth", *i.e.* the derivative in the error term is *not* more or less constant in the interval. This may seem like a rare occurrence, yet in discontinuous or otherwise non-smooth functions (*e.g.* Fig. 4.2), this type of problem can arise even for higher-degree quadrature rules.

The probability of such an unfortunate coincidence is rather small and decreases as the degree of the quadrature rules used in the difference increases. Another way of reducing this probability is to consider *more than one* derivative of the function, which is what the error estimators in the fourth category do rather explicitly: O'Hara and Smith (Section 2.8)

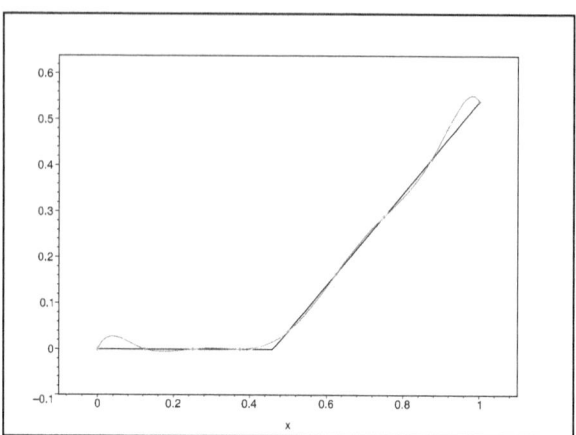

Figure 4.2: A non-smooth function (black) interpolated by $\tilde{f}_1(x)$ (green) using the nodes of $\mathsf{NC}_5^{(1)}[a,b]$ and $\tilde{f}_2(x)$ (red) using the nodes of $\mathsf{NC}_9^{(2)}[a,b]$.

effectively approximate three derivatives, Oliver (Section 2.11) computes the three highest-degree even Chebyshev coefficients and Berntsen and Espelid (Section 2.19) use a non-linear combination of higher-degree coefficients computed using null rules.

4.1.2   Non-Linear Error Estimators

Although most of the non-linear error estimators presented in Chapter 2 differ significantly in their approach, they all rely on the same basic principle, namely the assumption that, for any quadrature rule $Q^{(m)}[a,b]$, for sufficiently smooth $f(x)$ in the interval $x \in [a,b]$, the error can be written as

$$Q^{(m)}[a,b] - \int_a^b f(x)\,\mathrm{d}x \approx \kappa h^\alpha, \quad h = \frac{b-a}{m} \tag{4.1}$$

where κ depends on the basic quadrature rule Q and the higher derivatives of the integrand and α is the order of the error. In the most general case, Equation 4.1 has three unknowns, namely the actual integral $I = \int_a^b f(x)\,\mathrm{d}x$, the scaling κ and the order α of the error. In many cases, the order α is assumed to be the order of the quadrature rule, but in the presence of singularities or discontinuities, this is not always the case. The three unknowns may be resolved using three successive approximations:

$$Q^{(m)} = I + \kappa h^\alpha \tag{4.2}$$
$$Q^{(2m)} = I + \kappa h^\alpha 2^{-\alpha} \tag{4.3}$$
$$Q^{(4m)} = I + \kappa h^\alpha 4^{-\alpha} \tag{4.4}$$

We can subtract Equation 4.2 from Equation 4.3 to isolate the error term

$$Q^{(m)} - Q^{(2m)} = \kappa h^\alpha \left(1 - 2^{-\alpha} \right)$$
$$\kappa h^\alpha = \frac{Q^{(m)} - Q^{(2m)}}{1 - 2^{-\alpha}} = \frac{2^\alpha \left(Q^{(m)} - Q^{(2m)} \right)}{2^\alpha - 1}. \tag{4.5}$$

Re-inserting this expression into Equation 4.3, we obtain

$$Q^{(2m)} = I + \frac{2^\alpha \left(Q^{(m)} - Q^{(2m)} \right)}{2^\alpha - 1} 2^{-\alpha}$$
$$I = Q^{(2m)} - \frac{Q^{(m)} - Q^{(2m)}}{2^\alpha - 1} \tag{4.6}$$

which is the linear extrapolation used in the Romberg T-table (for even integer values of α) and also used by de Boor's CADRE (see Section 2.9, Equation 2.39), where the $Q^{(m)}$, $Q^{(2m)}$ and $Q^{(4m)}$ are the T-table entries $T_{\ell-2,i}$, $T_{\ell-1,i}$ and $T_{\ell,i}$ respectively, for an unknown α.

Inserting Equation 4.5 into Equation 4.3 and Equation 4.4 and taking the difference of the two, we can extract

$$Q^{(2m)} - Q^{(4m)} = \frac{2^\alpha \left(Q^{(m)} - Q^{(2m)} \right)}{2^\alpha - 1} \left(2^{-\alpha} - 4^{-\alpha} \right)$$
$$2^\alpha = \frac{Q^{(2m)} - Q^{(4m)}}{Q^{(m)} - Q^{(2m)}} \tag{4.7}$$

which is the ratio R_i used by de Boor (Equation 2.36) to approximate the order of the error ($2^{\alpha+1}$ therein).

Inserting both Equation 4.5 and Equation 4.7 into the last estimate,

Equation 4.4, we obtain

$$
\begin{aligned}
I &= Q^{(4m)} - \frac{Q^{(m)} - Q^{(2m)}}{1 - \frac{Q^{(2m)} - Q^{(4m)}}{Q^{(m)} - Q^{(2m)}}} \left(\frac{Q^{(2m)} - Q^{(4m)}}{Q^{(m)} - Q^{(2m)}} \right)^2 \\
&= Q^{(4m)} - \frac{\left(Q^{(m)} - Q^{(2m)}\right)^2}{Q^{(m)} - 2Q^{(2m)} + Q^{(4m)}} \frac{\left(Q^{(2m)} - Q^{(4m)}\right)^2}{\left(Q^{(m)} - Q^{(2m)}\right)^2} \\
&= Q^{(4m)} - \frac{\left(Q^{(2m)} - Q^{(4m)}\right)^2}{Q^{(m)} - 2Q^{(2m)} + Q^{(4m)}}
\end{aligned}
\tag{4.8}
$$

which is one step of the well-known Aitken Δ^2-process [1], for accelerating the convergence of a sequence of linearly converging estimates, which, if Equation 4.1 holds, is what the individual $Q^{(m)}[a, b]$ are.

The approach taken by Rowland and Varol (see Section 2.10) is almost identical, except that, instead of using the exact integral, they use

$$
\begin{aligned}
Q^{(m)} &= Q^{(2m)} + \kappa h^\alpha & (4.9) \\
Q^{(2m)} &= Q^{(4m)} + \kappa h^\alpha 2^{-\alpha} & (4.10) \\
Q^{(4m)} &= I + \kappa h^\alpha 4^{-\alpha} & (4.11)
\end{aligned}
$$

to solve for κh^α, $2^{-\alpha}$ and the exact integral I, resulting in their simpler error estimate (Equation 2.46).

In a similar vein, Laurie (see Section 2.16) uses the equations

$$
\begin{aligned}
Q_\alpha^{(1)} &= I + \kappa_\alpha (b - a)^{\alpha+2} \\
Q_\alpha^{(2)} &= I + \kappa_\alpha (b - a)^{\alpha+2} 2^{-(\alpha+2)} \\
Q_\beta^{(1)} &= I + \kappa_\beta (b - a)^{\beta+2} \\
Q_\beta^{(2)} &= I + \kappa_\beta (b - a)^{\beta+2} 2^{-(\beta+2)}
\end{aligned}
\tag{4.12}
$$

which are, however, under-determined, since there are 5 unknowns (κ_α, κ_β, α, β and I) yet only four equations. To get a bound on the equation, Laurie therefore adds the conditions in Equation 2.78 and Equation 2.79, obtaining the inequality in Equation 2.77 from which he constructs his error estimate.

Similarly, Favati, Lotti and Romani use the equations

$$
\begin{aligned}
Q_\alpha &= I + \kappa_\alpha (b-a)^{\alpha+2} \\
Q_\beta &= I + \kappa_\beta (b-a)^{\beta+2} \\
Q_\gamma &= I + \kappa_\gamma (b-a)^{\gamma+2} \\
Q_\delta &= I + \kappa_\delta (b-a)^{\delta+2}
\end{aligned}
\tag{4.13}
$$

which have 8 unknowns, and which can be solved together with the four conditions in Equation 2.92 to Equation 2.95.

Laurie and Venter's error estimator (see Section 2.10), is an exception: Although similar in form to that of Rowland and Varol, it does not follow the same approach, since the estimates

$$
\begin{aligned}
Q_1^{(1)} &= I + \kappa_1 (b-a)^3 \\
Q_3^{(1)} &= I + \kappa_3 (b-a)^5 \\
&\;\;\vdots \\
Q_{255}^{(1)} &= I + \kappa_{255}(b-a)^{257}
\end{aligned}
\tag{4.14}
$$

form a set of n equations in $n+1$ unknowns (I and the n different κ_i, assuming, for simplicity, that the actual order of the error is that of the quadrature rule) which can *not* be solved as above.

In summary, these methods, *i.e.* Romberg's method, the Aitken Δ^2-process and Rowland and Varol's extrapolation, take a sequence of initial estimates $Q^{(m)}$, $Q^{(2m)}$, $Q^{(4m)}$, ... and use them to create a sequence of *improved* estimates by removing the dominant error term as per Equation 4.1. These approaches can, of course, be re-applied to the resulting sequence, thus eliminating the next dominant error term, and so on. This is exactly what is done in the columns of the Romberg T-table and in successive re-applications of the Aitken Δ^2-process.

In all these methods, the error estimate is taken to be *the difference between the last estimate and the extrapolated value of the integral*. In the case of de Boor's CADRE, this is the difference between the last two entries in the bottom row of the modified T-table, and for Rowland and Varol (Section 2.10), Laurie (Section 2.16) and Favati, Lotti and Romani (Section 2.16), this is $Q^{(4m)} - I$, $Q_\alpha^{(2)} - I$ and $Q_\alpha - I$ respectively.

Instead of successively and iteratively removing the dominant term in the error, we could also simply model the error directly as the sum of several powers

$$Q^{(m)} - I \approx \kappa_1 h^{\alpha_1} + \kappa_2 h^{\alpha_2} + \cdots + \kappa_N h^{\alpha_N}, \quad h = \frac{b-a}{m} \quad (4.15)$$

Since this equation has $2N + 1$ unknowns (the n constants κ_i, the N exponents α_i and the exact integral I), we need $2N + 1$ estimates to solve for them:

$$
\begin{aligned}
Q^{(m)} &= I + \kappa_1 h^{\alpha_1} + \kappa_2 h^{\alpha_2} + \cdots + \kappa_N h^{\alpha_N} \quad (4.16)\\
Q^{(2m)} &= I + \kappa_1 h^{\alpha_1} 2^{-\alpha_1} + \kappa_2 h^{\alpha_2} 2^{-\alpha_2} + \cdots + \kappa_N h^{\alpha_N} 2^{-\alpha_N}\\
Q^{(4m)} &= I + \kappa_1 h^{\alpha_1} 2^{-2\alpha_1} + \kappa_2 h^{\alpha_2} 2^{-2\alpha_2} + \cdots + \kappa_N h^{\alpha_N} 2^{-2\alpha_N}\\
Q^{(8m)} &= I + \kappa_1 h^{\alpha_1} 2^{-3\alpha_1} + \kappa_2 h^{\alpha_2} 2^{-3\alpha_2} + \cdots + \kappa_N h^{\alpha_N} 2^{-3\alpha_N}\\
&\vdots\\
Q^{(2^{2N}m)} &= I + \kappa_1 h^{\alpha_1} 2^{-2n\alpha_1} + \kappa_2 h^{\alpha_2} 2^{-2n\alpha_2} + \cdots + \kappa_N h^{\alpha_N} 2^{-2N\alpha_N}
\end{aligned}
$$

This non-linear system of equations does not appear to be an easy thing to solve, yet in [50] Kahaner shows that, if we are only interested in I, this is *exactly* what the ϵ-Algorithm [102] does. For an even number of approximations $2N$, the algorithm computes the same approximation as in Equation 4.15, yet only over the first $N - 1$ terms, ignoring the first estimate $Q^{(m)}$.

Keeping Equation 4.15 in mind, de Doncker's error estimate (see Section 2.14, Equation 2.69) then reduces to

$$\varepsilon_i \approx 2 \left| \kappa_N h^{\alpha_N} \right|$$

for $N = \lfloor i/2 \rfloor$, assuming that, ideally, for all estimates the right-most even column of the epsilon-table was used.

Kahaner also shows that if such an expansion as per Equation 4.15 exists with more than one non-integer α_i, then the Romberg T-table and the Aitken Δ^2-process, applied iteratively, will insert spurious terms in the error expansion. Although both methods eventually converge to the correct result, they do so much more slowly than the ϵ-Algorithm.

In summary, we can say that all the error estimators presented in Chapter 2 assume that the error of a quadrature rule $Q^{(m)}[a, b]$ behaves as

$$Q^{(m)} - I \approx \kappa_1 h^{\alpha_1} + \kappa_2 h^{\alpha_2} + \cdots + \kappa_N h^{\alpha_N}, \quad h = \frac{b-a}{m}. \quad (4.17)$$

The unknowns in this equation ($I = \int_a^b f(x)\,\mathrm{d}x$, κ_i and α_i) can be solved approximately using several approximations $Q_n^{(m)}$.

If the exponents α_i are known or assumed to be known, the resulting system is a *linear* system of equations. This is what Romberg's method computes quite explicitly and what many of the linear error estimators in Section 4.1.1 compute implicitly. If the exponents α_i are *not* known, the resulting system of equations is *non-linear* and can therefore only be solved for non-linearly.

The non-linear methods discussed here are therefore a conceptual extension of the linear error estimators presented earlier. As such, they are subject to the same problem of the difference between two estimates being accidentally small in cases where the assumptions in Equation 4.1 or Equation 4.17 do not actually hold, as is the case for singular or discontinuous integrands. The different error estimation techniques in this section differ only in the depth N of the expansion and the use of additional constraints when the resulting system of equations is under-determined.

Although it is impossible to test if the assumptions on which these error estimates are based actually *hold*, it is possible to test if they are *plausible* This is done in most cases, yet as Espelid and Sørevik [25] show, these tests are not always adequate and their failure to detect a problem can lead to the catastrophic failures of the algorithm.

4.2   A New Error Estimator

Instead of constructing our error estimate from the difference between different estimates of the *integral* of the integrand, as is done in almost all the methods analyzed in Section 4.1, we will use the L_2-norm of the difference between the different *interpolations* of the integrand to construct:

$$\varepsilon = \sqrt{\int_{-1}^{1} \left[\hat{f}(x) - g_n(x) \right]^2 \mathrm{d}x} \quad (4.18)$$

where $\hat{f}(x)$ is the integrand $f(x)$ transformed from the interval $[a, b]$ to $[-1, 1]$

$$\hat{f}(x) = f\left(\frac{a+b}{2} + \frac{b-a}{2}x\right)$$

and $g_n(x)$ is the interpolation polynomial of degree n resulting from interpolating the integrand at the nodes of the quadrature rule:

$$g_n(x_i) = \hat{f}(x_i), \quad i = 0 \ldots n.$$

The error estimate in Equation 4.18 is an approximation of the integration error of the interpolant $g_n(x)$

$$Q_n[-1, 1] - \int_a^b \hat{f}(x)\,\mathrm{d}x = \int_{-1}^1 (g_n(x) - \hat{f}(x))\,\mathrm{d}x$$

and will only be zero if the interpolated integrand matches the integrand on the entire interval

$$g_n(x) = \hat{f}(x), \quad x \in [-1, 1].$$

In such a case, the integral will also be computed exactly. The error Equation 4.18 is therefore, assuming we can evaluate Equation 4.18 reliably, not susceptible to "accidentally small" values.

Since we don't know $f(x)$ explicitly, we cannot evaluate the right-hand side of Equation 4.18 exactly. In a first, naive approach, we could compute two interpolations $g_{n_1}^{(1)}(x)$ and $g_{n_2}^{(2)}(x)$ of different degree where $n_1 < n_2$. If we assume, similarly to Piessens' and Patterson's error estimates (see Section 2.12) in which the estimate from a higher-degree rule is used to estimate the error of a lower-degree rule, that $g_{n_2}^{(2)}(x)$ is a sufficiently precise approximation of the integrand

$$g_{n_2}^{(2)}(x) \approx f(x), \quad x \in [-1, 1] \tag{4.19}$$

then we can approximate the error of the interpolation $g_{n_1}^{(1)}(x)$ as

$$\left[\int_{-1}^1 \left(f(x) - g_{n_1}^{(1)}(x)\right)^2 \mathrm{d}x\right]^{1/2} \approx \left[\int_{-1}^1 \left(g_{n_2}^{(2)}(x) - g_{n_1}^{(1)}(x)\right)^2 \mathrm{d}x\right]^{1/2}$$

$$\approx \|\mathbf{c}^{(2)} - \mathbf{c}^{(1)}\|$$

where $c^{(1)}$ and $c^{(2)}$ are the vectors of the coefficients of $g_{n_1}^{(1)}(x)$ and $g_{n_2}^{(2)}(x)$ respectively and $c_i^{(1)} = 0$ where $i > n_1$. Our first, naive error estimate is hence

$$\varepsilon = (b - a)\|c^{(1)} - c^{(2)}\|. \tag{4.20}$$

This error estimate, however, is only valid for the lower-degree interpolation $g_{n_1}^{(1)}(x)$ and would over-estimate the error of the higher-degree interpolation $g_{n_2}^{(2)}(x)$ which we would use to compute the integral. For a more refined error estimate, we could consider the interpolation error

$$\hat{f}(x) - g_n(x) = \frac{\hat{f}^{(n+1)}(\xi_x)}{(n+1)!}\pi_{n+1}(x), \quad \xi_x \in [-1, 1] \tag{4.21}$$

for any n times continuously differentiable $f(x)$ where ξ_x depends on x and where $\pi_{n+1}(x)$ is the Newton polynomial over the $n+1$ nodes of the quadrature rule:

$$\pi_{n+1}(x) = \prod_{i=0}^{n}(x - x_i).$$

Taking the L_2-norm on both sides of Equation 4.21 we obtain

$$\varepsilon = \left[\int_{-1}^{1}\left(g_n(x) - \hat{f}(x)\right)^2 dx\right]^{1/2} =$$

$$\left[\int_{-1}^{1}\left(\frac{\hat{f}^{(n+1)}(\zeta_x)}{(n+1)!}\right)^2 \pi_{n+1}^2(x)\, dx\right]^{1/2}.$$

Since $\pi_{n+1}^2(x)$ is, by definition, positive for any x, we can apply the mean value theorem of integration and extract the derivative resulting in

$$\varepsilon = \left[\int_{-1}^{1}\left(g_n(x) - \hat{f}(x)\right)^2 dx\right]^{1/2} =$$

$$\left|\frac{\hat{f}^{(n+1)}(\xi)}{(n+1)!}\right|\left[\int_{-1}^{1}\pi_{n+1}^2(x)\, dx\right]^{1/2}, \quad \xi \in [-1, 1]. \tag{4.22}$$

If we represent the polynomial $\pi_{n+1}(x)$ analogously to $g_n(x)$, as

$$\pi_{n+1}(x) = \sum_{k=0}^{n+1} b_k p_k(x), \qquad (4.23)$$

then we can compute its L_2-norm as

$$\sqrt{\int_{-1}^{1} \pi_{n+1}^2(x)\,\mathrm{d}x} = \|\mathbf{b}\|$$

where $\mathbf{b}$ is the vector of the $n+2$ coefficients[2] b_k. Therefore, the terms on the right-hand side of Equation 4.22, only the $(n+1)$st derivative of the integrand is unknown.

Given two interpolations of the integrand, $g_n^{(1)}(x)$ and $g_n^{(2)}(x)$, of the same degree yet not over the same set of nodes, if we assume that the derivative $f^{(n+1)}(\xi)$ is constant for $\xi \in [a,b]$[3], we can extract the unknown derivative as follows:

$$
\begin{aligned}
g_n^{(1)}(x) - g_n^{(2)}(x) &= \hat{f}(x) + \frac{\hat{f}^{(n+1)}(\xi)}{(n+1)!}\pi_{n+1}^{(1)}(x) - \hat{f}(x) - \frac{\hat{f}^{(n+1)}(\xi)}{(n+1)!}\pi_{n+1}^{(2)}(x) \\
&= \frac{\hat{f}^{(n+1)}(\xi)}{(n+1)!}\left(\pi_{n+1}^{(1)}(x) - \pi_{n+1}^{(2)}(x)\right) \qquad (4.24)
\end{aligned}
$$

where $\pi_{n+1}^{(1)}(x)$ and $\pi_{n+1}^{(2)}(x)$ are the $(n+1)$st Newton polynomials over the $n+1$ nodes of $g_n^{(1)}(x)$ and $g_n^{(2)}(x)$ respectively. Taking the L_2-norm on both sides of (4.24), we obtain

$$
\sqrt{\int_{-1}^{1}\left[g_n^{(1)}(x) - g_n^{(2)}(x)\right]^2\,\mathrm{d}x} = \left|\frac{\hat{f}^{(n+1)}(\xi)}{(n+1)!}\right|\sqrt{\int_{-1}^{1}\left[\pi_{n+1}^{(1)}(x) - \pi_{n+1}^{(2)}(x)\right]^2\,\mathrm{d}x}
$$

$$
\left\|\mathbf{c}^{(1)} - \mathbf{c}^{(2)}\right\| = \left|\frac{\hat{f}^{(n+1)}(\xi)}{(n+1)!}\right|\left\|\mathbf{b}^{(1)} - \mathbf{b}^{(2)}\right\|
$$

$$
\left|\frac{\hat{f}^{(n+1)}(\xi)}{(n+1)!}\right| = \frac{\left\|\mathbf{c}^{(1)} - \mathbf{c}^{(2)}\right\|}{\left\|\mathbf{b}^{(1)} - \mathbf{b}^{(2)}\right\|} \qquad (4.25)
$$

---

[2] In [46], Higham shows how the coefficients of a Newton-like polynomial can be computed relative to any orthogonal base.

[3] This assumption is a stronger form of the "sufficiently smooth" condition, which we will use only to construct the error estimator.

from which we can construct an error estimate for either interpolation

$$\sqrt{\int_{-1}^{1} \left[g_n^{(1)}(x) - \hat{f}(x) \right]^2 \, dx} = \frac{\left\| \mathbf{c}^{(1)} - \mathbf{c}^{(2)} \right\|}{\left\| \mathbf{b}^{(1)} - \mathbf{b}^{(2)} \right\|} \left\| \mathbf{b}^{(1)} \right\|$$

$$\sqrt{\int_{-1}^{1} \left[g_n^{(2)}(x) - \hat{f}(x) \right]^2 \, dx} = \frac{\left\| \mathbf{c}^{(1)} - \mathbf{c}^{(2)} \right\|}{\left\| \mathbf{b}^{(1)} - \mathbf{b}^{(2)} \right\|} \left\| \mathbf{b}^{(2)} \right\|. \qquad (4.26)$$

For the interpolation $g_n^{(1)}(x)$, this gives us the refined error estimate

$$\varepsilon_2 = (b - a) \frac{\left\| \mathbf{c}^{(1)} - \mathbf{c}^{(2)} \right\|}{\left\| \mathbf{b}^{(1)} - \mathbf{b}^{(2)} \right\|} \left\| \mathbf{b}^{(1)} \right\| \qquad (4.27)$$

Note that for this estimate, we have made the assumption that the nth derivative is constant. We can't verify this directly, but we can verify if our computed $\left| \frac{f^{(n+1)}(\xi)}{(n+1)!} \right|$ (4.25) actually satisfies (4.21) for the nodes of the first interpolation by testing

$$\left| g_n^{(2)}(x_i) - \hat{f}(x_i) \right| \le \vartheta_1 \left| \frac{f^{(n)}(\xi)}{n!} \right| \left| \pi_{n+1}^{(2)}(x_i) \right|, \quad i = 1 \ldots n \qquad (4.28)$$

where the x_i are the nodes of the interpolation $g_n^{(1)}(x)$ and the value $\vartheta_1 \ge 1$ is an arbitrary relaxation parameter. If this condition is violated for any of the x_i, then we use, as in the naive error estimate (4.20), the un-scaled error estimate

$$\varepsilon = (b - a) \left\| \mathbf{c}^{(1)} - \mathbf{c}^{(2)} \right\|. \qquad (4.29)$$

In practice, we can implement this error estimator in a recursive adaptive quadrature by first computing the $n + 1$ coefficients c_k of $g_n(x)$ in the interval $[a, b]$. We also need the $n+2$ coefficients b_k of the $(n+1)$st Newton polynomial over the nodes of the basic quadrature rule. If the same rule is used throughout the algorithm, then the b_k can be pre-computed.

For the first interval, no error estimate is computed. The interval is bisected and for the recursion on the left half of $[a, b]$, we compute

$$\mathbf{c}^{\text{old}} = \mathbf{T}^{(\ell)} \mathbf{c}, \quad \mathbf{b}^{\text{old}} = \mathbf{T}^{(\ell)} \mathbf{b}. \qquad (4.30)$$

Note that to compute $\mathbf{b}^{\text{old}}$ we would actually need to extend $\mathbf{T}^{(\ell)}$ and, since $\mathbf{b}^{\text{old}}$ and $\mathbf{b}$ are not in the same interval, we have to scale the co-efficients of $\mathbf{b}^{\text{old}}$ by 2^{n+2} so that Equation 4.21 holds for $g_n^{(2)}(x)$ in the sub-interval.

Inside the left sub-interval $[a, (a + b)/2]$, we then evaluate the new co-efficients $\mathbf{c}$ and $\mathbf{b}$. Note that if the base quadrature rule is the same throughout the algorithm, this is the same vector $\mathbf{b}$ we used above.

Given the old and new coefficients, we then compute the error estimate

$$\varepsilon_2 = (b - a)\frac{\|\mathbf{c} - \mathbf{c}^{\text{old}}\|}{\|\mathbf{b} - \mathbf{b}^{\text{old}}\|}\|\mathbf{b}\| \tag{4.31}$$

which is then either compared to a local tolerance τ_k in the case of a purely recursive algorithm along the lines of Algorithm 1, or summed to compute a global error estimate in the case of a heap-based algorithm as per Algorithm 2.

4.3   Numerical Tests

In the following, we will compare the performance of some of the error estimation techniques discussed in Section 4.1, including the new error estimator presented in Section 4.2.

4.3.1   Methodology

Whereas other authors [8, 48, 49, 66, 87, 53] have focused on comparing different algorithms as a whole, using sets of functions chosen to best represent typical integrands, we will focus here only on the error estimators and integrands chosen such that they specifically should or should not cause the error estimator to fail.

For these test functions we will not consider the usual metrics of efficiency, *i.e.* number of function evaluations required for a given accuracy, but the number of *false negatives* and *false positives* for each error estimator when integrating functions which it should or should not integrate correctly, respectively.

We define a *false positive* as a returned error estimate which is *below* the required tolerance when the actual error is *above* the latter. Likewise, a *false negative* is a returned error estimate which is *above* the required tolerance when the actual error is *below* the latter.

In practical terms, false negatives are a sign that the error estimator is overly cautious and continues to refine the interval even though the required tolerance would already have been achieved. False positives, however, may cause the algorithm to fail completely: if the actual error in a sub-interval is larger than the global tolerance, no amount of excess precision in the other intervals will fix it and the result will be incorrect, save an identical false positive elsewhere of opposite sign.

The test integrands, along with an explanation of why they were chosen, are:

1. $p_n(x)$: The Chebyshev polynomial of degree n in the interval $[-\alpha, \beta]$, where α and β are chosen randomly in $(0, 1]$ and n is the degree of the quadrature rule for which the error estimate is computed[4]. The polynomial is shifted by $+1$ to avoid an integral of zero.

   This function should, by design, be integrated exactly by all quadrature routines and hence any error estimate larger than the required tolerance is a false negative.

2. $p_{n+1}(x)$: Same as the function above, yet one degree above the degree of the quadrature rule.

   Although this integrand is, by design, beyond the degree of the quadrature rule, the error term (*i.e.* the $n + 1$st derivative) is constant and can be extrapolated reliably from the difference of two quadratures of degree n, *e.g.* as is done implicitly in SQUANK (see Section 2.7, Equation 2.20) or explicitly in Ninomiya's error estimator (see Section 2.15, Equation 2.73), thus providing an exact error estimate.

3. $p_{n+2}(x)$: Same as the function above, yet two degrees above the degree of the quadrature rule.

---

[4] For error estimates computed from the difference of two quadrature rules of different degree, the degree of the quadrature rule of lower degree is used since although the result rule of higher degree is effectively used for the returned integrand, the error estimate is usually understood to be that of the lower-degree rule.

By design, the $n + 1$st derivative is linear in x and changes sign at the origin, meaning that any attempt to extrapolate that derivative from two estimates of equal degree may fail, causing an incorrect error estimate unless this is specifically tested for, *e.g.* using the noise detection in SQUANK (see Section 2.7) or the decision process in CADRE (see Section 2.9).

4. $d_k(x)$: A function with a discontinuity at $x = \alpha$ in the kth derivative, where α is chosen randomly in the interval of integration $[-1, 1]$ for $k = 0, 1$ and 2:

$$d_0(x) = \begin{cases} 0 & x < \alpha \\ 1 & \text{otherwise} \end{cases} \tag{4.32}$$

$$d_1(x) = \max\{0, x - \alpha\} \tag{4.33}$$

$$d_2(x) = (\max\{0, x - \alpha\})^2 \tag{4.34}$$

Since all quadrature rules considered herein are interpolatory in nature and these integrands can not be reliably interpolated, these functions will only be correctly integrated by chance. The only exception is CADRE (see Section 2.9), which attempts to detect jump discontinuities explicitly.

5. $s(x)$: A function with an integrable singularity at $x = \alpha$, where α is chosen randomly in $(-1, 1)$:

$$s(x) = |x - \alpha|^{-1/2}$$

As with the previous set of functions, this function can not be reliably interpolated and an interpolatory quadrature rule will produce a correct result only by chance. The only exception is again CADRE, which treats such singularities explicitly when detected (see Section 2.9, Equation 2.38) in cases where α is near the edges of the domain.

These functions were tested for $10\,000$ realizations of the random parameters α and β for each of the relative tolerances $\tau = 10^{-1}$, 10^{-3}, 10^{-6}, 10^{-9} and 10^{-12}. Since most error estimators use absolute tolerances, the tolerance was set to the respective fraction of the integral. The following error estimators were implemented in Matlab [67] (see Appendix A) and tested:

1. Kuncir's error estimate (Section 2.2, Equation 2.1), where $n = 3$ is the degree of the composite Simpson rules used,

2. McKeeman's 1962 error estimate (Section 2.3, Equation 2.5), where $n = 3$ is the degree of the composite Simpson rules used,

3. McKeeman's 1963 variable-order error estimate with $n = 7$ (Section 2.5, Equation 2.14), which is also the degree of the composite Newton-Cotes rules used,

4. Gallaher's error estimate (Section 2.6, Equation 2.17), where $n = 2$ is the degree of the trapezoidal rule used,

5. Lyness' error estimate as implemented in SQUANK (Section 2.7, Equation 2.18), where $n = 3$ is the degree of the Simpson's rules used. Note that two levels of recursion are evaluated to check for noise and, if it is detected, 2τ is returned,

6. O'Hara and Smith's error estimate (Section 2.8, Equation 2.22, Equation 2.23 and Equation 2.24), which, if implemented as described in their original manuscript, the algorithm only rarely gets past the first error estimate (Equation 2.22) and is therefore not suited for testing the more interesting, third error estimate (Equation 2.24). For this reason, the algorithm was implemented *without* the initial error estimate and hence $n = 6$,

7. Oliver's error estimate (Section 2.11, Equation 2.55), where $n = 9$ is the degree of the second-last Clenshaw-Curtis quadrature used and the first error estimate below tolerance is returned or 2τ if the interval is to be subdivided,

8. Piessens' error estimate (Section 2.12, Equation 2.57), where $n = 19$ is the degree of the 10-point Gauss quadrature used,

9. Patterson's error estimate (Section 2.12, Equation 2.58), where $n = 191$ is the degree of the second-last quadrature rule used and the first error estimate below tolerance is returned or 2τ if the error estimate for 255 nodes is still larger than tolerance. Since Patterson's original algorithm is known to be prone to rounding and truncation errors, the improved implementation by Krogh and Snyder [52] was used,

10. Garribba *et al.* 's SNIFF error estimate (Section 2.13, Equation 2.64) using a 6-point Gauss-Legendre quadrature rule with $n = 11$ as suggested in [35],

11. Ninomiya's error estimate (Section 2.15, Equation 2.73) using the 9-point quadrature rule with $n = 9$ and the 11-point error estimator,

12. QUADPACK's QAG error estimator (Section 2.17, Equation 2.97) using the 10-point Gauss quadrature rule with $n = 19$ and its 21-point Kronrod extension,

13. Berntsen and Espelid's error estimate (Section 2.18, Equation 2.99) using the 21-point Gauss quadrature rule and the 20-point interpolatory quadrature rule with $n = 19$,

14. Berntsen and Espelid's null-rule error estimate (Section 2.19, Equation 2.107 and Equation 2.108) using, as a basic quadrature rule, the 21-point Clenshaw-Curtis quadrature rule (the 21-point Gauss quadrature rule was also tried but left out since it produced worse results, *i.e.* more false positives) with $n = 21$ and values $K = 3$, $r_{\text{critical}} = 1/4$ and $\alpha = 1/2$.

15. Gander and Gautschi's error estimate (Section 2.20, Equation 2.114) using the 4-point Gauss-Lobatto quadrature rule with $n = 5$ and its 7-point Kronrod extension,

16. De Boor's CADRE error estimate (Section 2.9, Equation 2.37) where the entire decision process was considered with a maximum T-table depth of 10 and thus $n = 17$

17. Rowland and Varol's error estimate (Section 2.10, Equation 2.46) with $m = 1$ and $n = 3$ from the underlying Simpson's rule,

18. Venter and Laurie's error estimate (Section 2.10, Equation 2.50) where $n = 19$ is the degree of the second-last stratified quadrature rule and where the algorithm decides to subdivide an error estimate of 2τ is returned,

19. Laurie's sharper error estimate (Section 2.16, Equation 2.77) using the 10-point Gauss quadrature rule with $n = 19$ and its 21-point Kronrod extension for the two rules Q_β and Q_α respectively, as suggested by Laurie himself in [57],

20. Favati, Lotti and Romani's error estimate (Section 2.16, Equation 2.96), using as quadrature rules the 7 and 10 point Gauss quadrature rules as well as their 15 and 21 point Kronrod extensions with $n = 21$ being the degree of the 15 point Kronrod extension,

21. Kahaner and Stoer's error estimate (Section 2.14, Equation 2.70) using, for simplicity, the full trapezoidal sums (*i.e.* bisecting all intervals) with a maximum depth of $i = 7$ and using the ϵ-Algorithm to extrapolate the integral, where error estimate was computed from the difference between the final estimate and the previous two estimates, the lower of which has degree $n = 5$,

22. The new, trivial error estimate (Section 4.2, Equation 4.20) using the nodes of the $n = n_1 = 11$ and $n_2 = 21$-point Clenshaw-Curtis quadrature rules to compute the two interpolations $g_{n_1}^{(1)}(x)$ and $g_{n_2}^{(2)}(x)$ respectively.

23. The new, more refined error estimate (Section 4.2, Equation 4.27) using the nodes of an 11-point Clenshaw-Curtis quadrature rule with $n = 10$ and one level of recursion to obtain $\mathbf{c}^{\text{old}}$, as well as 1.1 for the constant ϑ_1 in Equation 4.28.

Not all error estimators described in the previous sections were included in this test. The error estimate of McKeeman and Tesler (Section 2.4) is identical to McKeeman's first error estimate (Section 2.3), differing only in how the local tolerance is adjusted after bisection. Forsythe, Malcolm and Moler's error estimate (Section 2.7) is also not included since it is only an extension of Lyness' error estimate. Favati *et al.* 's error estimate using recursively monotone stable rules (Section 2.17) is, in essence, identical to that of QUADPACK's QAG. The different variants of Berntsen and Espelid's null-rule based error estimate (Section 2.19) were condensed into the version used here since the error estimation procedure is in essence identical. Finally, de Doncker's error estimate (Section 2.14) was not included since it works pretty much in the same way as Kahaner and Stoer's variant with the ϵ-Algorithm, yet requiring much fewer nodes. The local error estimates are provided by the same error estimate as QUADPACK's QAG, which is included in the test, and hence will be as reliable as the latter.

Function	$\tau = 10^{-1}$	$\tau = 10^{-3}$	$\tau = 10^{-6}$	$\tau = 10^{-9}$	$\tau = 10^{-12}$
$p_n(x)$	100 (0/0)	100 (0/0)	100 (0/0)	100 (0/0)	100 (0/0)
$p_{n+1}(x)$	65.67 (0/34.33)	8.49 (0/21.14)	0.38 (0/0.76)	0.01 (0/0.02)	0 (0/0.01)
$p_{n+2}(x)$	51.07 (0/48.93)	8.44 (0/15.77)	0.50 (0/1.07)	0.03 (0/0.11)	0.02 (0/0)
$d_0(x)$	16.58 (0/22.27)	0 (0/0.35)	0 (0/0)	0 (0/0)	0 (0/0)
$d_1(x)$	44.92 (0/29.98)	0.73 (0/1.59)	0 (0/0)	0 (0/0)	0 (0/0)
$d_2(x)$	54.30 (0/22.66)	5.74 (0/7.16)	0.22 (0/0.12)	0.01 (0/0)	0 (0/0)
$s(x)$	0 (33.05/17.11)	0 (0.42/0.20)	0 (0/0)	0 (0/0)	0 (0/0)

Table 4.1: Results for Kuncir's 1962 error estimate.

Function	$\tau = 10^{-1}$	$\tau = 10^{-3}$	$\tau = 10^{-6}$	$\tau = 10^{-9}$	$\tau = 10^{-12}$
$p_n(x)$	100 (0/0)	100 (0/0)	100 (0/0)	100 (0/0)	100 (0/0)
$p_{n+1}(x)$	64.19 (0/35.81)	8.06 (0/51.21)	0.26 (0/2.19)	0.01 (0/0.05)	0 (0/0)
$p_{n+2}(x)$	49.14 (0/50.86)	8.29 (0/37.35)	0.45 (0/2.51)	0.04 (0/0.15)	0.01 (0/0)
$d_0(x)$	0 (0/53.59)	0 (0/0.74)	0 (0/0)	0 (0/0)	0 (0/0)
$d_1(x)$	47.48 (0/35.05)	0.90 (0/4.73)	0 (0/0.01)	0 (0/0)	0 (0/0)
$d_2(x)$	54.57 (0/30.24)	5.52 (0/28.66)	0.17 (0/0.49)	0.02 (0/0.02)	0 (0/0)
$s(x)$	0.03 (21.61/23.13)	0 (0.21/0.21)	0 (0/0)	0 (0/0)	0 (0/0)

Table 4.2: Results for McKeeman's 1962 error estimate.

4.3.2 Results

The results of the tests described in Section 4.3.1 are shown in Tables 4.1 to 4.23. For each integrand and tolerance, the percentage of correct integrations is given (*i.e.* the error estimate and the actual error are both below the required tolerance), as well as, in brackets, the percentage of false positives and false negatives respectively.

The results for Kuncir's error estimate (Section 2.2) are summarized in Table 4.1. Despite the low degree of the quadrature rule and its simplicity, the error estimator performs rather well. The function $p_n(x)$ is integrated reliably as would be expected. For the higher-degree $p_{n+1}(x)$, the error estimate is exact save a constant factor of 15, resulting in about as many false negatives as correct predictions with no false positives. Surprisingly enough, the function $p_{n+2}(x)$ is also handled correctly, despite the zero in the $n + 1$st derivative, causing no false positives. The discontinuous functions do not generate any false positives either and the error estimator returns positive results only for the lowest tolerances. Finally, the singularity is not detected for $\tau = 10^{-1}$ in more than a third of the cases, returning a false positive.

The results for McKeeman's 1962 error estimate (Section 2.3) are summarized in Table 4.2. As with Kuncir's error estimate, with which it is

Function	$\tau = 10^{-1}$	$\tau = 10^{-3}$	$\tau = 10^{-6}$	$\tau = 10^{-9}$	$\tau = 10^{-12}$
$p_n(x)$	100 (0/0)	100 (0/0)	100 (0/0)	100 (0/0)	100 (0/0)
$p_{n+1}(x)$	93.20 (0/6.80)	41.73 (0/58.27)	8.07 (0/91.93)	1.54 (0/62.01)	0.31 (0/11.38)
$p_{n+2}(x)$	87.13 (0/12.87)	33.28 (0/66.72)	6.97 (0/93.03)	1.25 (0/48.18)	0.46 (0/9.99)
$d_0(x)$	56.53 (1.03/37.73)	0.26 (3.56/4.84)	0 (2.12/0)	0 (2.11/0)	0 (1.87/0)
$d_1(x)$	84.52 (0/12.91)	4.65 (0.08/72.32)	0 (0/0.23)	0 (0/0)	0 (0/0)
$d_2(x)$	82.86 (0/14.69)	30.27 (0/58.70)	0.32 (0/14.20)	0.01 (0/0.25)	0 (0/0)
$s(x)$	35.25 (0.39/55.61)	0 (0.44/0.56)	0 (0/0)	0 (0/0)	0 (0/0)

Table 4.3: Results for McKeeman's 1963 error estimate.

Function	$\tau = 10^{-1}$	$\tau = 10^{-3}$	$\tau = 10^{-6}$	$\tau = 10^{-9}$	$\tau = 10^{-12}$
$p_n(x)$	100 (0/0)	100 (0/0)	100 (0/0)	100 (0/0)	100 (0/0)
$p_{n+1}(x)$	0 (0/0)	0 (0/0)	0 (0/0)	0 (0/0)	0 (0/0)
$p_{n+2}(x)$	1.21 (0/47.06)	0.07 (0/2.45)	0 (0/0.02)	0 (0/0)	0 (0/0)
$d_0(x)$	18.32 (81.68/0)	0.28 (99.72/0)	0 (100/0)	0 (100/0)	0 (100/0)
$d_1(x)$	23.77 (76.23/0)	3.20 (96.80/0)	0.06 (99.94/0)	0 (100/0)	0 (100/0)
$d_2(x)$	0 (0/0)	0 (0/0)	0 (0/0)	0 (0/0)	0 (0/0)
$s(x)$	0 (0/19.09)	0 (0/0.18)	0 (0/0)	0 (0/0)	0 (0/0)

Table 4.4: Results for Gallaher's 1967 error estimate.

almost identical, it integrates the function $p_n(x)$ exactly and returns no false positives for the functions $p_{n+1}(x)$ and $p_{n+2}(x)$, yet the rate of false negatives is slightly higher due to the larger constant factor of 80 vs. 15 in the error term (Equation 2.7). As with Kuncir's error estimator the discontinuities are handled properly and the singularity is not detected for low tolerances, albeit less often due to more conservative scaling factor of the error. The same conclusions are also valid for McKeeman's 1963 error estimate (Section 2.5, Table 4.3), which is even more conservative – the error is over-estimated by a factor of 5 764 800 – resulting in an even larger number of false negatives for the polynomials, and, interestingly enough, some false positives for the discontinuity $d_0(x)$.

The results for Gallaher's 1967 error estimate (Section 2.6) are summarized in Table 4.4. The results for $p_n(x)$ and $p_{n+1}(x)$ are correct as expected, despite the odd form of the second derivative (Equation 2.17), which does not compute the error term exactly. Although the algorithm correctly fails on the singularity $s(x)$, it does not detect the first two discontinuities $d_0(x)$ and $d_1(x)$, returning false positives in almost all cases. As for the second degree discontinuity $d_2(x)$, it is beyond the degree of the integrator and the error is thus predicted correctly. If the correct expression (Equation 2.16) with the same constant scaling of 14.1 is used, better results are obtained (not shown), yet with some (ca. 10%) false

Function	$\tau = 10^{-1}$	$\tau = 10^{-3}$	$\tau = 10^{-6}$	$\tau = 10^{-9}$	$\tau = 10^{-12}$
$p_n(x)$	100 (0/0)	100 (0/0)	100 (0/0)	100 (0/0)	100 (0/0)
$p_{n+1}(x)$	100 (0/0)	83.59 (0/16.41)	4.36 (0/95.64)	0.24 (0/99.76)	0 (0/100)
$p_{n+2}(x)$	100 (0/0)	54.66 (0/45.34)	3.41 (0/96.59)	0.28 (0/99.72)	0.01 (0/99.99)
$d_0(x)$	63.20 (33.91/0)	0 (0/0.57)	0 (0/0)	0 (0/0)	0 (0/0)
$d_1(x)$	87.11 (7.66/0)	6.89 (30.14/1.46)	0 (0.04/0)	0 (0/0)	0 (0/0)
$d_2(x)$	85.57 (6.07/0)	42.96 (16.35/0.85)	0.50 (1.39/0.01)	0.03 (0.04/0)	0 (0/0)
$s(x)$	26.90 (70.92/0)	0 (4.84/0.32)	0 (0/0)	0 (0/0)	0 (0/0)

Table 4.5: Results for Lyness' 1969 error estimate.

Function	$\tau = 10^{-1}$	$\tau = 10^{-3}$	$\tau = 10^{-6}$	$\tau = 10^{-9}$	$\tau = 10^{-12}$
$p_n(x)$	100 (0/0)	86.69 (0/13.31)	7.32 (0/92.68)	0.21 (0/99.79)	0 (0/100)
$p_{n+1}(x)$	100 (0/0)	62.89 (0/37.11)	8.88 (0/91.12)	1.09 (0/98.91)	0.19 (0/99.81)
$p_{n+2}(x)$	99.72 (0/0.28)	49.87 (0/50.13)	7.24 (0/46.45)	0.70 (0/8.81)	0.10 (0/1.70)
$d_0(x)$	24.79 (0/47.57)	0 (0/1.16)	0 (0/0)	0 (0/0)	0 (0/0)
$d_1(x)$	76.59 (0.80/15.09)	4.27 (1.36/21.91)	0 (0/0.03)	0 (0/0)	0 (0/0)
$d_2(x)$	83.95 (0/11.31)	24.42 (0/40.68)	0.29 (0/1.80)	0.01 (0/0.01)	0 (0/0)
$s(x)$	5.02 (35.95/34.85)	0 (0/0.29)	0 (0/0)	0 (0/0)	0 (0/0)

Table 4.6: Results for O'Hara and Smith's 1969 error estimate.

positives for $d_2(x)$.

The results for Lyness' 1969 error estimate (Section 2.7) are summarized in Table 4.5. For the polynomials $p_{n+1}(x)$ and $p_{n+2}(x)$ the number of false negatives is almost 100% since although the error for the compound Simpson's rule of degree 3 is computed, the 5-point Newton-Cotes quadrature of degree 5 is returned as an the integral. The quadratures are therefore exact, but this is not recognized by the error estimate. Although the algorithm implicitly tests for noise or odd behavior in the integrand, it does not reliably catch the first or second degree discontinuities $d_1(x)$ and $d_2(x)$ or the singularity $s(x)$, producing a large number of false positives for moderate tolerances. Although this error estimator is in essence identical to that of Kuncir, it's exact scaling of the difference of the quadratures by $1/15$ is what makes it fail more often for the discontinuities.

The results for O'Hara and Smith's 1969 error estimate (Section 2.8) are summarized in Table 4.6. The algorithm drastically over-predicts the error for the polynomial $p_n(x)$ since it's estimate uses the last coefficient of the integrand, which is large by construction, to infer the magnitude of the following coefficients, which are zero by construction. The errors for $p_{n+1}(x)$ and $p_{n+2}(x)$ are similarly over-predicted. The interpolations of the discontinuities $d_0(x)$, $d_1(x)$ and $d_2(x)$ often have large higher-degree

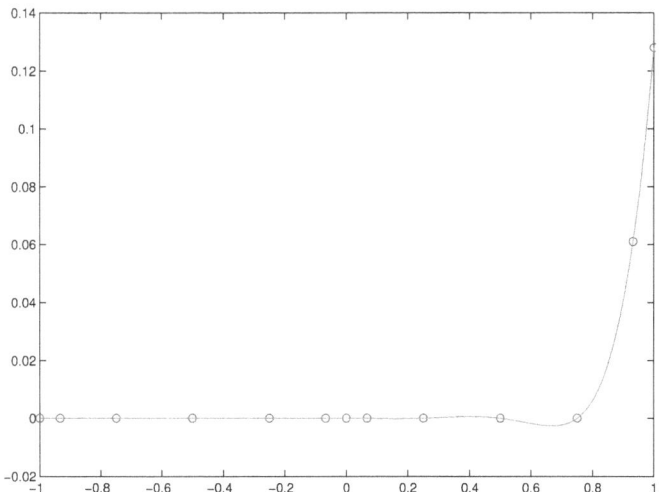

Figure 4.3: The polynomial interpolated by O'Hara and Smith's error estimate (red) for the discontinuous integrand $d_1(x)$ (green).

coefficients which are caught by the error estimate except for some cases in $d_1(x)$ in which the higher-degree Chebyshev coefficients are small and thus the interpolated integrand looks smooth, causing the error estimate to under-predict the error (see Fig. 4.3). Similarly, the error for singularity $s(x)$ is mis-predicted whenever the interpolation is *too* smooth (see Fig. 4.4).

The results for Oliver's 1972 error estimate (Section 2.11) are summarized in Table 4.7. Similarly to O'Hara and Smith's error estimator, the algorithm mis-predicts the errors for all three polynomials $p_n(x)$, $p_{n+1}(x)$

Function	$\tau = 10^{-1}$	$\tau = 10^{-3}$	$\tau = 10^{-6}$	$\tau = 10^{-9}$	$\tau = 10^{-12}$
$p_n(x)$	65.69 (2.40/31.91)	22.20 (0.25/77.55)	8.67 (0/91.33)	2.77 (0/97.23)	0.72 (0/99.28)
$p_{n+1}(x)$	55.07 (3.87/41.06)	18.34 (0.22/69.25)	6.03 (0/21.70)	1.18 (0/5.57)	0.23 (0/1.13)
$p_{n+2}(x)$	49.62 (5.79/44.59)	14.93 (0.30/64.31)	5.72 (0/18.04)	1.52 (0/5.15)	0.50 (0/1.58)
$d_0(x)$	20.44 (0/35.08)	0 (0/0.64)	0 (0/0)	0 (0/0)	0 (0/0)
$d_1(x)$	71.27 (0.86/18.23)	3.60 (6.96/10.86)	0 (0/0.03)	0 (0/0)	0 (0/0)
$d_2(x)$	78.09 (0/16.14)	23.55 (5.33/18.77)	0.35 (0/0.90)	0.01 (0/0.03)	0 (0/0)
$s(x)$	2.06 (66.71/15.27)	0 (0.60/0.23)	0 (0/0)	0 (0/0)	0 (0/0)

Table 4.7: Results for Oliver's 1972 error estimate.

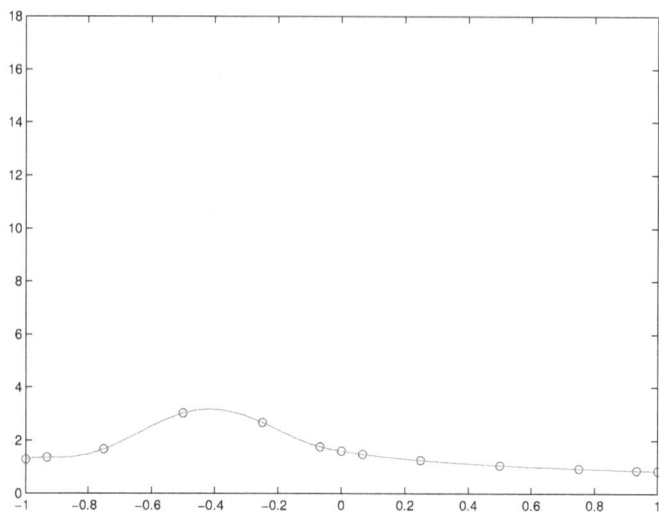

Figure 4.4: The polynomial interpolated by O'Hara and Smith's error estimate (red) for the singular integrand $s(x)$ (green).

and $p_{n+2}(x)$, due to the large higher-degree coefficients of the integrands. The false positives are cases where the doubly-adaptive algorithm exited after incorrectly predicting the error with a lower-order rule. This is also true for the discontinuities $d_0(x)$, $d_1(x)$ and $d_2(x)$, which are detected well by the higher-oder rules since the higher-degree Chebyshev coefficients become relatively large, yet fail when the error is mis-predicted by the lower-degree rules. The algorithm fails when integrating the singularity $s(x)$, since the coefficients of the interpolation often decay smoothly, misleading it to believe the integrand itself is smooth, similarly to O'Hara and Smith's error estimate (see Fig. 4.4).

The results for Piessens' 1973 error estimate (Section 2.12) are summarized in Table 4.8. As would be expected, the error estimate for the polynomial $p_n(x)$ is exact. For the higher-degree polynomials $p_{n+1}(x)$ and $p_{n+2}(x)$ the high number of false negatives is due to the higher-degree Kronrod extension being used as an approximation to the integral although the error is estimated for its underlying lower-degree Gauss rule. The error for the discontinuities $d_0(x)$, $d_1(x)$ and $d_2(x)$ is approximated

Function	$\tau = 10^{-1}$	$\tau = 10^{-3}$	$\tau = 10^{-6}$	$\tau = 10^{-9}$	$\tau = 10^{-12}$
$p_n(x)$	100 (0/0)	100 (0/0)	100 (0/0)	100 (0/0)	100 (0/0)
$p_{n+1}(x)$	98.02 (0/1.98)	83.76 (0/16.24)	52.05 (0/47.95)	25.96 (0/74.04)	12.69 (0/87.31)
$p_{n+2}(x)$	96.05 (0/3.95)	76.82 (0/23.18)	42.95 (0/57.05)	21.48 (0/78.52)	11.60 (0/88.40)
$d_0(x)$	63.54 (0.34/22.58)	0.14 (0.27/2.21)	0.01 (0.46/0)	0 (0.39/0)	0 (0.44/0)
$d_1(x)$	94.56 (0.21/4.21)	18.19 (3.98/28.56)	0.14 (0.32/0.16)	0 (0.46/0)	0 (0.40/0)
$d_2(x)$	95.70 (0.28/3.53)	55.06 (0.40/27.79)	1.18 (0.56/4.51)	0.11 (0.44/0.01)	0.01 (0.46/0)
$s(x)$	38.89 (25.21/23.75)	0 (1.05/0.36)	0 (0/0)	0 (0/0)	0 (0/0)

Table 4.8: Results for Piessens' 1973 error estimate.

Function	$\tau = 10^{-1}$	$\tau = 10^{-3}$	$\tau = 10^{-6}$	$\tau = 10^{-9}$	$\tau = 10^{-12}$
$p_n(x)$	42.37 (57.63/0)	97.69 (2.31/0)	100 (0/0)	100 (0/0)	100 (0/0)
$p_{n+1}(x)$	43.24 (56.76/0)	97.55 (2.45/0)	100 (0/0)	100 (0/0)	100 (0/0)
$p_{n+2}(x)$	44.01 (55.99/0)	97.90 (2.10/0)	100 (0/0)	100 (0/0)	100 (0/0)
$d_0(x)$	87.12 (12.88/0)	0.10 (22.66/12.78)	0 (22.27/0.01)	0 (22.60/0)	0 (22.68/0)
$d_1(x)$	88.99 (11.01/0)	71.84 (28.16/0)	0.14 (24.07/4.98)	0 (21.89/0)	0.01 (22.60/0)
$d_2(x)$	87.33 (12.67/0)	88.28 (11.72/0)	56.91 (12.46/25.98)	0.72 (13.13/4.70)	0.01 (12.85/0)
$s(x)$	20.31 (79.56/0.06)	0 (9.73/1.16)	0 (0/0)	0 (0/0)	0 (0/0)

Table 4.9: Results for Patterson's 1973 error estimate.

reliably in most cases due to the high degree of the quadrature rule, which more reliably interpolates the function. This, however, fails now and then due to accidentally small differences between the Gauss and Gauss-Kronrod rules, producing false positives. The singularity $s(x)$ produces many false positives at $\tau = 10^{-1}$ for the same reason (see Fig. 4.5).

Patterson's 1973 error estimate (Section 2.12), for which the results are summarized in Table 4.9, suffers from the same problems as Oliver's when using cascading error estimates. For the polynomials $p_n(x)$, $p_{n+1}(x)$ and $p_{n+2}(x)$ and in some cases the singularity $s(x)$ at moderate tolerances, the error estimator mis-predicts the errors using the lower-order quadrature rules, causing a large number of false positives. The exact predictions for higher tolerances for $p_{n+1}(x)$ and $p_{n+2}(x)$ are due to the relatively small contribution to the integral of the error term at such high degree ($n = 191$). The high number of false positives for the discontinuities $d_0(x)$, $d_1(x)$ and $d_2(x)$ are caused when the discontinuity at α is outside of the first 7 quadrature nodes, making it practically undetectable. The remaining false positives when integrating the singularity $s(x)$ are due to the same accidentally small differences observed for Piessens' 1973 error estimate (see Fig. 4.5).

The results for Garribba et al.'s 1980 error estimate (Section 2.13) are summarized in Table 4.10. The error estimate is extremely precise for

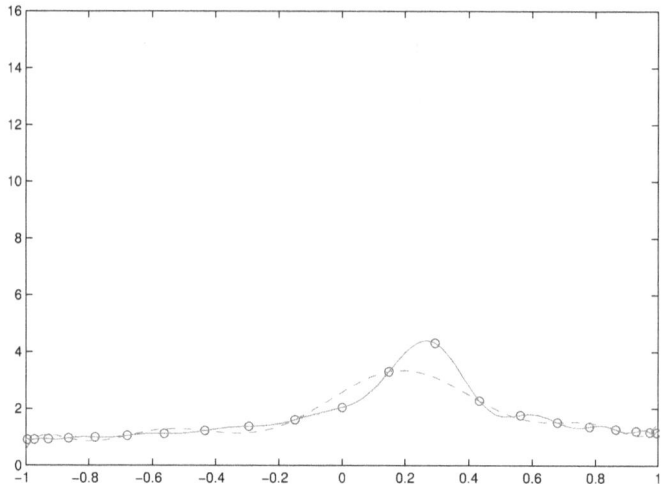

Figure 4.5: The integrands assumed by the Gauss (red dashed) and Gauss-Kronrod (red solid) quadratures in Piessens' 1973 error estimate for the singular integrand $s(x)$ (green).

Function	$\tau = 10^{-1}$	$\tau = 10^{-3}$	$\tau = 10^{-6}$	$\tau = 10^{-9}$	$\tau = 10^{-12}$
$p_n(x)$	100 (0/0)	100 (0/0)	100 (0/0)	100 (0/0)	100 (0/0)
$p_{n+1}(x)$	100 (0/0)	100 (0/0)	74.94 (0/0)	26.16 (0/0)	8.70 (0/0)
$p_{n+2}(x)$	100 (0/0)	100 (0/0)	63.34 (0/0)	22.54 (0/0)	7.47 (0/0)
$d_0(x)$	74.63 (25.37/0)	1.26 (98.74/0)	0 (6.69/0.01)	0 (7.12/0)	0 (6.72/0)
$d_1(x)$	93.09 (6.91/0)	29.45 (70.55/0)	0.15 (29.92/0.09)	0.02 (3.75/0)	0 (3.22/0)
$d_2(x)$	94.97 (5.03/0)	66.17 (33.83/0)	2.20 (56.55/0.03)	0.18 (4.42/0.14)	0.02 (3.63/0)
$s(x)$	40.88 (59.12/0)	0.23 (99.30/0)	0 (0.94/0)	0 (0.01/0)	0 (0/0)

Table 4.10: Results for Garribba *et al.* 's 1978 error estimate.

Function	$\tau = 10^{-1}$	$\tau = 10^{-3}$	$\tau = 10^{-6}$	$\tau = 10^{-9}$	$\tau = 10^{-12}$
$p_n(x)$	100 (0/0)	100 (0/0)	100 (0/0)	100 (0/0)	100 (0/0)
$p_{n+1}(x)$	90.28 (0/9.72)	49.13 (0/50.87)	11.90 (0/88.10)	2.37 (0/97.63)	0.40 (0/99.60)
$p_{n+2}(x)$	83.59 (0/16.41)	38.58 (0/61.42)	10.94 (0/89.06)	3.06 (0/96.94)	0.85 (0/99.15)
$d_0(x)$	23.84 (0.52/42.09)	0 (0/1.02)	0 (0/0)	0 (0/0)	0 (0/0)
$d_1(x)$	73.27 (0.32/19.85)	4.36 (1.79/14.75)	0 (0/0.01)	0 (0/0)	0 (0/0)
$d_2(x)$	82.87 (0/9.71)	21.25 (0.12/28.62)	0.48 (0.08/0.15)	0.01 (0.01/0)	0 (0/0)
$s(x)$	4.30 (33.41/29.49)	0 (0.53/0.27)	0 (0/0)	0 (0/0)	0 (0/0)

Table 4.11: Results for Ninomiya's 1980 error estimate.

Function	$\tau = 10^{-1}$	$\tau = 10^{-3}$	$\tau = 10^{-6}$	$\tau = 10^{-9}$	$\tau = 10^{-12}$
$p_n(x)$	100 (0/0)	100 (0/0)	100 (0/0)	100 (0/0)	100 (0/0)
$p_{n+1}(x)$	84.04 (0/15.96)	70.01 (0/29.99)	47.75 (0/52.25)	30.61 (0/69.39)	18.19 (0/81.81)
$p_{n+2}(x)$	76.68 (0/23.32)	60.87 (0/39.13)	38.91 (0/61.09)	25.60 (0/74.40)	16.22 (0/83.78)
$d_0(x)$	6.04 (0.32/79.64)	0.11 (0.29/2.06)	0 (0.49/0)	0 (0.45/0)	0 (0.38/0)
$d_1(x)$	22.50 (0.21/76.36)	1.43 (0.35/44.96)	0.12 (0.45/0.22)	0.01 (0.52/0)	0 (0.44/0)
$d_2(x)$	57.99 (0.18/41.19)	15.36 (0.28/67.99)	0.79 (0.30/5.23)	0.09 (0.34/0)	0.03 (0.48/0)
$s(x)$	0.26 (0.54/62.29)	0 (0.03/0.35)	0 (0/0)	0 (0/0)	0 (0/0)

Table 4.12: Results for Piessens *et al.*'s 1983 error estimate.

the polynomials $p_n(x)$, $p_{n+1}(x)$ and $p_{n+2}(x)$, generating no false negatives. The discontinuities $d_0(x)$, $d_1(x)$ and $d_2(x)$ and the singularity $s(x)$, however, are not well detected at all, resulting in many false positives over all tolerances since the underlying assumption of smoothness (Equation 2.59) is not valid for these functions.

The results for Ninomiya's 1980 error estimate (Section 2.15) are summarized in Table 4.11. The error for the polynomial $p_n(x)$ is correctly approximated. For $p_{n+1}(x)$, the large number of false positives is due to the subtraction of the (exact) error estimate from the approximation to the integral, thus returning an approximation of higher degree than the quadrature itself, and $p_{n+2}(x)$ causes no special difficulties. The discontinuities $d_0(x)$, $d_1(x)$ and $d_2(x)$ and the singularity $s(x)$ are integrated with a low number of false positives, even for moderate tolerances, due to the approximation of the derivative being "accidentally small".

Piessens *et al.*'s 1983 error estimate (Section 2.17) does a very good job over all functions (Table 4.12). The error estimate generates a high number of false negatives for the polynomials $p_{n+1}(x)$ and $p_{n+2}(x)$ since the quadrature rule used to approximate the integral is several degrees more exact then that for which the returned error estimate is computed. The few false positives are due to the error estimate's scaling of the error, causing it to under-predict the actual error and to cases where

Function	$\tau = 10^{-1}$	$\tau = 10^{-3}$	$\tau = 10^{-6}$	$\tau = 10^{-9}$	$\tau = 10^{-12}$
$p_n(x)$	100 (0/0)	100 (0/0)	100 (0/0)	100 (0/0)	100 (0/0)
$p_{n+1}(x)$	98.62 (0/1.38)	85.42 (0/14.58)	52.85 (0/47.15)	26.09 (0/73.91)	13.52 (0/86.48)
$p_{n+2}(x)$	97.03 (0/2.97)	78.10 (0/21.90)	44.73 (0/55.27)	22.57 (0/77.43)	12.23 (0/87.77)
$d_0(x)$	66.48 (0.35/20.07)	0.14 (0.63/2.06)	0 (0.61/0)	0 (0.59/0)	0 (0.57/0)
$d_1(x)$	96.70 (0.32/2.08)	20.73 (3.70/26.64)	0.08 (0.53/0.08)	0 (0.72/0)	0 (0.72/0)
$d_2(x)$	97.66 (0.34/1.52)	56.70 (0.75/26.47)	2.05 (0.64/4.47)	0.13 (0.48/0.01)	0.01 (0.54/0)
$s(x)$	44.02 (24.34/19.57)	0 (1.27/0.39)	0 (0.01/0)	0 (0/0)	0 (0/0)

Table 4.13: Results for Berntsen's 1984 error estimate.

Function	$\tau = 10^{-1}$	$\tau = 10^{-3}$	$\tau = 10^{-6}$	$\tau = 10^{-9}$	$\tau = 10^{-12}$
$p_n(x)$	51.98 (0/48.02)	23.69 (0/76.31)	8.15 (0/91.85)	2.56 (0/97.44)	0.97 (0/99.03)
$p_{n+1}(x)$	48.42 (0/51.58)	21.97 (0/78.03)	7.24 (0/78.24)	2.13 (0/54.11)	0.84 (0/29.48)
$p_{n+2}(x)$	43.89 (0/56.11)	20.23 (0/79.77)	6.77 (0/71.77)	2.34 (0/45.22)	0.73 (0/26.05)
$d_0(x)$	53.45 (0/31.20)	0 (0/1.86)	0 (0/0)	0 (0/0)	0 (0/0)
$d_1(x)$	85.10 (0/13.32)	3.76 (0/41.23)	0 (0/0.26)	0 (0/0)	0 (0/0)
$d_2(x)$	90.18 (0/8.94)	34.92 (0/47.13)	0.27 (0/5.23)	0 (0/0.11)	0 (0/0)
$s(x)$	13.03 (28.88/45.80)	0 (0/0.34)	0 (0/0)	0 (0/0)	0 (0/0)

Table 4.14: Results for Berntsen and Espelid's 1991 error estimate.

the discontinuity at α was outside of the open nodes of the quadrature rule. The error estimates for the discontinuities $d_0(x)$, $d_1(x)$ and $d_2(x)$ generate almost no false positives and the integration of the singularity $s(x)$ at $\tau = 10^{-1}$ generates less than 1% false positives for the same reasons as Piessens' 1973 error estimate (see Fig. 4.5).

Berntsen and Espelid's error estimate (Section 2.18), for which the results are summarized in Table 4.13, does as well on the polynomials $p_n(x)$, $p_{n+1}(s)$ and $p_{n+2}(x)$ as Piessens *et al.*'s, after which it is modelled. On the discontinuities $d_0(x)$, $d_1(x)$ and $d_2(x)$, however, it generates more false positives and does poorly on the singularity $s(x)$ for moderate tolerances. Since the two quadratures from which the error is computed differ by only a single node, the difference in the resulting interpolations will be small for functions which can *not* be interpolated reliably (see Fig. 4.6).

Berntsen and Espelid's null-rule error estimate (Section 2.19), for which the results are summarized in Table 4.14, suffers from the same problems as O'Hara and Smith's and Oliver's error estimates for the polynomial $p_n(x)$: Although the integration is exact, the coefficients $\tilde{c}_i$ increase towards $i = n$, leading the algorithm to believe that the $n + 1$st coefficient will be large when it is, in fact, zero. This pessimistic approximation, however, works to their advantage for the remaining polynomials $p_{n+1}(x)$ and $p_{n+2}(x)$ as well as for the discontinuities $d_0(x)$, $d_1(x)$ and $d_2(x)$. The

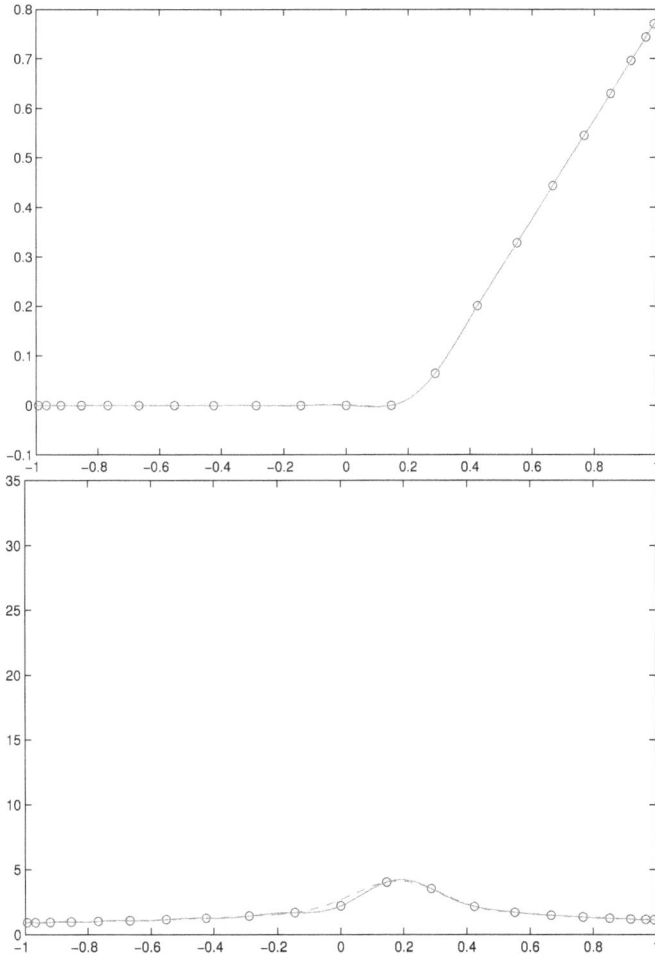

Figure 4.6: The integrands assumed by the quadrature rules $Q_{20}[a, b]$ (red dashed) and $G_{21}[a, b]$ (red solid) in Berntsen's 1984 error estimate for the discontinuous and singular integrands $d_1(x)$ and $s(x)$ respectively (green).

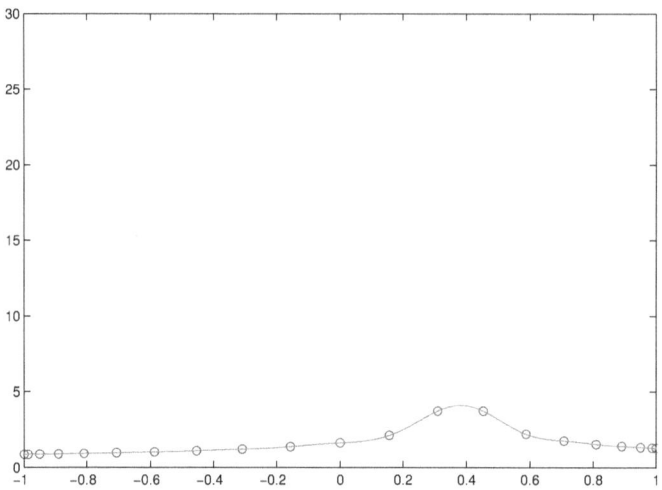

Figure 4.7: The integrand assumed by Berntsen and Espelid's 1991 error estimate (red) for the singular integrand $s(x)$ (green).

algorithm mis-predicts the error for the singularity $s(x)$ for the same reason as Oliver's algorithm, namely that the coefficients of the polynomial interpolation decrease smoothly, falsely indicating convergence (see Fig. 4.7).

The results for Gander and Gautschi's error estimate (Section 2.20) are summarized in Table 4.15. The algorithm performs well on the polynomials $p_n(x)$, $p_{n+1}(x)$ and $p_{n+2}(x)$, generating no false positives. The high number of false negatives for $p_{n+1}(x)$ and $p_{n+2}(x)$ are due to the higher degree of the estimate effectively returned. The error estimation

Function	$\tau = 10^{-1}$	$\tau = 10^{-3}$	$\tau = 10^{-6}$	$\tau = 10^{-9}$	$\tau = 10^{-12}$
$p_n(x)$	100 (0/0)	100 (0/0)	100 (0/0)	100 (0/0)	100 (0/0)
$p_{n+1}(x)$	80.08 (0/19.92)	17.69 (0/82.31)	0.56 (0/99.44)	0 (0/100)	0 (0/99.99)
$p_{n+2}(x)$	68.15 (0/31.85)	17.88 (0/82.12)	2.46 (0/97.54)	0.33 (0/99.67)	0.08 (0/99.92)
$d_0(x)$	10.33 (0/39.32)	0 (0/0.59)	0 (0/0)	0 (0/0)	0 (0/0)
$d_1(x)$	63.43 (2.32/23.63)	0.70 (1.33/9.97)	0 (0/0)	0 (0/0)	0 (0/0)
$d_2(x)$	68.98 (0/19.77)	8.69 (0.03/25.79)	0.31 (0/0.13)	0.02 (0/0.01)	0 (0/0)
$s(x)$	0 (44.15/22.67)	0 (0.50/0.22)	0 (0/0)	0 (0/0)	0 (0/0)

Table 4.15: Results for Gander and Gautschi's 2001 error estimate.

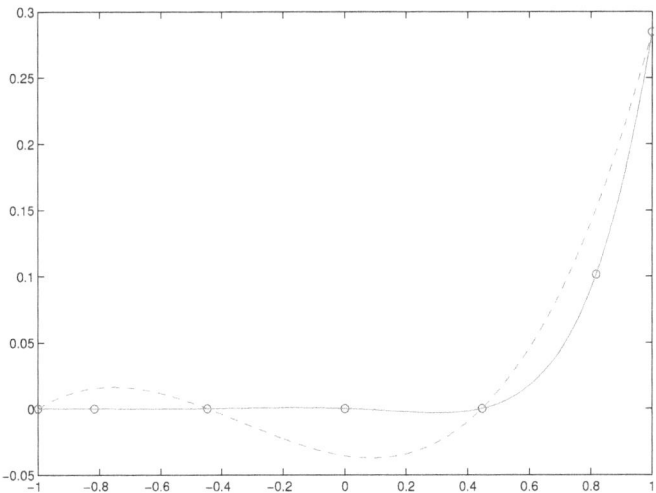

Figure 4.8: The integrands assumed by the Gauss-Lobatto (red dashed) and Gauss-Kronrod (red solid) quadrature rules in Gander and Gautschi's 2001 error estimate over the discontinuous integrand $d_1(x)$ (green).

returns some false positives for the discontinuities $d_0(x)$, $d_1(x)$ and $d_2(x)$, as well as for the singularity $s(x)$, due to the difference between the two quadrature rules used being "accidentally small" (*e.g.* Fig. 4.8).

The results for de Boor's error estimate (Section 2.9), as shown in Table 4.16, are somewhat mixed. As with other doubly adaptive routines, it often fails to integrate the polynomial $p_n(x)$. Specifically, De Boor's error estimate fails because the ratio R_i (Equation 2.36) is not close enough to 4 in the first few rows of the T-table and the algorithm therefore assumes that the integrand is not smooth enough. The false positives in all polynomials $p_n(x)$, $p_{n+1}(x)$ and $p_{n+2}(x)$ are caused by bad (*i.e.* "accidentally small") error estimates in the lower-degree rules. Otherwise, the jump discontinuity $d_0(x)$ is often detected as such and (correctly) treated explicitly. The algorithm, however, also treats $d_1(x)$ in the same way since the ratios R_i also converge to ≈ 2, resulting in a large number of false positives, since the error estimate assumes a jump in the 0th derivative. The third discontinuity $d_2(x)$ is only occasionally treated as a jump. In most cases it is treated as a normal, smooth function or in some cases as

Function	$\tau = 10^{-1}$	$\tau = 10^{-3}$	$\tau = 10^{-6}$	$\tau = 10^{-9}$	$\tau = 10^{-12}$
$p_n(x)$	25.73 (0.81/1.25)	24.94 (1.90/0.04)	23.78 (1.23/0)	23.40 (1.23/0)	23.26 (0.89/0)
$p_{n+1}(x)$	24.15 (1.04/1.38)	23.20 (2.32/0.09)	21.77 (0.76/0)	21.16 (1.33/0)	20.79 (1.05/0)
$p_{n+2}(x)$	21.80 (1.63/1.39)	20.72 (1.95/0.08)	19.54 (0.82/0)	19.18 (1.46/0)	18.92 (1.13/0)
$d_0(x)$	0 (0/57.95)	0 (0/0.75)	0 (0/0)	0 (0/0)	0 (0/0)
$d_1(x)$	67.83 (0/4.92)	3.83 (1.02/4.31)	0 (0.32/0.02)	0 (0.04/0)	0 (0.02/0)
$d_2(x)$	48.85 (0/0)	36.59 (12.60/0)	2.87 (32.78/0)	0.47 (35.38/0)	0.09 (33.75/0)
$s(x)$	1.14 (3.54/0.71)	0 (4.18/0)	0 (0.03/0)	0 (0/0)	0 (0/0)

Table 4.16: Results for De Boor's 1971 error estimate.

Function	$\tau = 10^{-1}$	$\tau = 10^{-3}$	$\tau = 10^{-6}$	$\tau = 10^{-9}$	$\tau = 10^{-12}$
$p_n(x)$	100 (0/0)	100 (0/0)	100 (0/0)	100 (0/0)	100 (0/0)
$p_{n+1}(x)$	100 (0/0)	83.58 (0.94/0)	3.98 (0.14/0)	0.08 (0/0)	0.01 (0/0)
$p_{n+2}(x)$	100 (0/0)	71.14 (1.57/0)	4.63 (0.06/0)	0.36 (0/0)	0.02 (0/0)
$d_0(x)$	40.85 (8.33/23.61)	0 (0/0.90)	0 (0/0)	0 (0/0)	0 (0/0)
$d_1(x)$	80.23 (0.58/7.32)	4.65 (12.12/5.35)	0 (0.53/0)	0 (0.01/0)	0 (0/0)
$d_2(x)$	81.89 (0/7.15)	36.63 (11.34/14.60)	0.82 (2.54/0.12)	0.03 (0.21/0.01)	0 (0.01/0)
$s(x)$	5.85 (43.68/22.25)	0 (7.42/0.24)	0 (0.22/0)	0 (0/0)	0 (0/0)

Table 4.17: Results for Rowland and Varol's 1972 error estimate.

a singularity, resulting in a large number of false positives. For the singularity $s(x)$, De Boor's special treatment for integrable singularities rarely kicks in and the algorithm usually aborts since the successive approximations do not seem to converge (*i.e.* the ratios R_i are not ≈ 4). When the singularity is detected, the extrapolation often returns an erroneous approximation of the error.

Rowland and Varol's error estimate (Section 2.10), for which the results are shown in Table 4.17, generates some false positives for the polynomials $p_{n+1}(x)$ and $p_{n+2}(x)$ where the error estimate was off by a small fraction due to "accidentally small" estimates by the lowest-degree rule. The discontinuities $d_0(x)$, $d_1(x)$ and $d_2(x)$ and the singularity $s(x)$ are not well detected since the integration error does not decay smoothly enough, *i.e.* the condition in Equation 2.45 is not valid, causing the error extrapolation to fail, resulting in a number of false positives for moderate tolerances. Note that better results may be achieved using larger values for m making the observed problems more rare, yet not resolving them completely.

As with other doubly-adaptive quadratures, Venter and Laurie's error estimate (Section 2.10), for which the results are summarized in Table 4.18, fails to integrate $p_n(x)$ correctly due to bad error estimates using the lower-degree rules, resulting in either bad values of hint (Equa-

Function	$\tau = 10^{-1}$	$\tau = 10^{-3}$	$\tau = 10^{-6}$	$\tau = 10^{-9}$	$\tau = 10^{-12}$
$p_n(x)$	3.83 (6.17/36.23)	2.44 (4.18/6.88)	2.12 (0.05/1.57)	0.40 (2.17/0.02)	0 (1.69/0)
$p_{n+1}(x)$	3.81 (5.76/35.27)	2.09 (3.89/5.97)	2.02 (0.14/1.58)	0.69 (1.66/0.01)	0.02 (1.36/0)
$p_{n+2}(x)$	3.44 (5.72/33.32)	2.28 (4.05/6.18)	1.87 (0.19/1.39)	0.86 (1.28/0.03)	0 (1.54/0)
$d_0(x)$	0 (0/47.96)	0 (0/0.58)	0 (0/0)	0 (0/0)	0 (0/0)
$d_1(x)$	25.86 (0/63.46)	0.29 (13.26/12.85)	0 (0.48/0.10)	0 (0.02/0.01)	0 (0/0)
$d_2(x)$	64.84 (0/26.22)	34.51 (19.83/9.32)	1.65 (12.77/1.02)	0 (2.70/0.01)	0 (1.53/0)
$s(x)$	0 (9.71/25.61)	0 (6.81/0.34)	0 (0.22/0)	0 (0.01/0)	0 (0/0)

Table 4.18: Results for Venter and Laurie's 2002 error estimate.

Function	$\tau = 10^{-1}$	$\tau = 10^{-3}$	$\tau = 10^{-6}$	$\tau = 10^{-9}$	$\tau = 10^{-12}$
$p_n(x)$	100 (0/0)	100 (0/0)	100 (0/0)	100 (0/0)	100 (0/0)
$p_{n+1}(x)$	100 (0/0)	100 (0/0)	100 (0/0)	100 (0/0)	100 (0/0)
$p_{n+2}(x)$	100 (0/0)	100 (0/0)	100 (0/0)	100 (0/0)	100 (0/0)
$d_0(x)$	30.26 (0.09/62.46)	0.12 (0.09/3.93)	0 (0.18/0.01)	0 (0.20/0)	0 (0.24/0)
$d_1(x)$	36.78 (0.07/62.75)	24.67 (3.78/48.51)	0.25 (1.14/0.55)	0 (0.41/0.01)	0 (0.46/0)
$d_2(x)$	44.81 (0.11/54.70)	40.21 (0.94/51.18)	3.52 (4.74/15.25)	0.14 (0.13/0.16)	0.03 (0.32/0)
$s(x)$	25.01 (0.06/64.82)	0 (4.52/0.52)	0 (0.03/0)	0 (0/0)	0 (0/0)

Table 4.19: Results for Laurie's 1983 error estimate.

tion 2.51) resulting in false negatives, or, less often, false convergence of the lower-degree rules used in the cascading error estimate, resulting in false positives. The use of **hint** guards the error estimate from making false assumptions on the discontinuity $d_0(x)$, yet not for $d_1(x)$ and $d_2(x)$ due to bogus convergence caused by "accidentally small" differences in the lower-degree rules, thus resulting in a large number of false positives. This works better, however, for the singularity $s(x)$, generating only few false positives.

The results for Laurie's error estimate (Section 2.16) are shown in Table 4.19. The error estimate is exact for the three polynomials $p_n(x)$, $p_{n+1}(x)$ and $p_{n+2}(x)$: even though the degree of the polynomials $p_{n+1}(x)$ and $p_{n+2}(x)$ is higher than the second-highest degree rule, the error of the highest-degree rule is correctly extrapolated as the polynomials are sufficiently smooth. Although these errors are themselves far below the required tolerance, they cause the original algorithm to fail when testing for the conditions in Equation 2.78 and Equation 2.80, where the difference between two otherwise exact quadrature rules becomes non-zero due to rounding and cancellation errors. This was partially amended by computing a somewhat arbitrary noise level

$$\varepsilon_{\mathsf{arb}} = 10\varepsilon_{\mathsf{mach}}(b-a)\max_i\{|f(x_i)|\} \qquad (4.35)$$

Function	$\tau = 10^{-1}$	$\tau = 10^{-3}$	$\tau = 10^{-6}$	$\tau = 10^{-9}$	$\tau = 10^{-12}$
$p_n(x)$	100 (0/0)	100 (0/0)	100 (0/0)	100 (0/0)	100 (0/0)
$p_{n+1}(x)$	100 (0/0)	100 (0/0)	100 (0/0)	100 (0/0)	100 (0/0)
$p_{n+2}(x)$	100 (0/0)	100 (0/0)	100 (0/0)	100 (0/0)	100 (0/0)
$d_0(x)$	72.73 (3.65/13.16)	0.21 (6.68/1.68)	0 (0.23/0)	0 (0.24/0)	0 (0.18/0)
$d_1(x)$	86.79 (0/12.20)	27.21 (9.03/18.60)	0.10 (0.48/0.22)	0.02 (0.25/0)	0 (0.27/0)
$d_2(x)$	94.28 (0/4.64)	68.74 (3.34/13.73)	2.49 (2.80/3.70)	0.03 (0.10/0.03)	0 (0.13/0)
$s(x)$	41.45 (24.22/20.66)	0 (2.72/0.34)	0 (0.01/0)	0 (0/0)	0 (0/0)

Table 4.20: Results for Favati *et al.* 's 1991 error estimate.

and truncating the moduli of the quadrature rules below it to zero. The discontinuities $d_0(x)$, $d_1(x)$ and $d_2(x)$ are not always detected since the condition in Equation 2.80 holds in some cases where the necessary condition in Equation 2.79 does not, resulting in some false positives over all tolerances and the error of the singularity $s(x)$ is, in most cases, extrapolated correctly.

The results for the error estimate of Favati *et al.* (Section 2.16) are shown in Table 4.20. This algorithm has a problem similar to that of Laurie's error estimate for integrands that are integrated exactly, *i.e.* the difference between two higher-order rules becomes zero or contains only noise, making the estimate fail. In their tests in [26], Favati *et al.* suggest avoiding this problem by aborting and bisecting the interval whenever $|Q_\alpha - Q_\beta|$, $|Q_\alpha - Q_\gamma|$ or $|Q_\alpha - Q_\delta|$ are below the required tolerance. This, however, would causes the algorithm to fail to converge for *any* integrand which can be integrated exactly by Q_β, which was not the case for any of their test integrands. This condition was therefore replaced with returning the result for Q_α if the error estimate $|Q_\alpha - Q_\beta|$ is below the noise level ε_{arb} defined in Equation 4.35. For the polynomials $p_n(x)$, $p_{n+1}(x)$ and $p_{n+2}(x)$, the noise exceeded this value only rarely, causing the occasional false negatives in these functions. The discontinuities $d_0(x)$, $d_1(x)$ and $d_2(x)$ and the singularity $s(x)$ were not always well detected since the conditions in Equation 2.92 to Equation 2.94 are not satisfied, resulting in a moderate number of false positives.

The result for Kahaner and Stoer's error estimate (Section 2.14), which are shown in Table 4.21, are somewhat mixed. The error estimator does a good job for the polynomials $p_n(x)$ The large number of false negatives for $p_{n+1}(x)$ and $p_{n+2}(x)$ is due to the algorithm returning a higher-degree estimate that the one for which the error estimate is computed. The results for the discontinuities, however, are not very encouraging. For

Function	$\tau = 10^{-1}$	$\tau = 10^{-3}$	$\tau = 10^{-6}$	$\tau = 10^{-9}$	$\tau = 10^{-12}$
$p_n(x)$	100 (0/0)	100 (0/0)	100 (0/0)	100 (0/0)	100 (0/0)
$p_{n+1}(x)$	100 (0/0)	99.85 (0/0.15)	37.83 (0/62.17)	1.31 (0/98.69)	0.05 (0/99.95)
$p_{n+2}(x)$	99.99 (0/0.01)	99.64 (0/0.36)	37.87 (0/62.13)	4.11 (0/95.89)	0.59 (0/99.35)
$d_0(x)$	57.55 (42.45/0)	1.22 (98.78/0)	0.01 (99.99/0)	0 (100/0)	0 (100/0)
$d_1(x)$	85.34 (13.84/0.22)	30.37 (53.23/11.73)	0.61 (67.36/0.41)	0 (68.25/0)	0 (67.71/0)
$d_2(x)$	87.84 (11.25/0.01)	78.42 (10.90/9.58)	5.96 (15.69/10.90)	0.40 (16.59/0.27)	0.02 (17.24/0.01)
$s(x)$	23.71 (3.03/35.37)	0 (0.30/0.68)	0 (0/0)	0 (0/0)	0 (0/0)

Table 4.21: Results for Kahaner and Stoer's 1983 error estimate.

Function	$\tau = 10^{-1}$	$\tau = 10^{-3}$	$\tau = 10^{-6}$	$\tau = 10^{-9}$	$\tau = 10^{-12}$
$p_n(x)$	100 (0/0)	100 (0/0)	100 (0/0)	100 (0/0)	100 (0/0)
$p_{n+1}(x)$	89.78 (0/10.22)	52.10 (0/47.90)	14.80 (0/85.20)	4.06 (0/95.94)	1.12 (0/98.88)
$p_{n+2}(x)$	81.73 (0/18.27)	40.76 (0/59.24)	12.22 (0/87.78)	4.52 (0/95.48)	1.34 (0/98.66)
$d_0(x)$	0 (0/84.09)	0 (0/2.31)	0 (0/0)	0 (0/0)	0 (0/0)
$d_1(x)$	66.03 (0/32.46)	0.34 (0/44.30)	0 (0/0.28)	0 (0/0)	0 (0/0)
$d_2(x)$	76.67 (0/22.50)	16.19 (0/65.95)	0.16 (0/5.34)	0.01 (0/0.12)	0 (0/0)
$s(x)$	0 (0/59.16)	0 (0/0.39)	0 (0/0)	0 (0/0)	0 (0/0)

Table 4.22: Results for the trivial error estimate in Equation 4.20.

the functions $d_0(x)$ and $d_1(x)$, the error estimator fails more often than not, resulting in many false positives. These false positives are due to false convergence in the ϵ-Algorithm, whenever the difference between two estimates was accidentally small. The results for the singularity $s(x)$ are comparable to other methods.

The results for the *new* error estimates (Equation 4.20 and Equation 4.31) described in Section 4.2 are summarized in Tables 4.22 and 4.23. In both estimators, the polynomials $p_n(x)$, $p_{n+1}(x)$ and $p_{n+2}(x)$ don't cause any problems, except that the errors for the two higher-degree polynomials tend to be over-estimated as the computed L_2-norm is a somewhat pessimistic estimate of the integration error. The discontinuities $d_0(x)$, $d_1(x)$ and $d_2(x)$ are correctly detected and in many cases integrated correctly

Function	$\tau = 10^{-1}$	$\tau = 10^{-3}$	$\tau = 10^{-6}$	$\tau = 10^{-9}$	$\tau = 10^{-12}$
$p_n(x)$	100 (0/0)	100 (0/0)	100 (0/0)	100 (0/0)	100 (0/0)
$p_{n+1}(x)$	100 (0/0)	100 (0/0)	58.76 (0/41.24)	17.49 (0/82.51)	5.15 (0/94.85)
$p_{n+2}(x)$	83.30 (0/16.70)	58.78 (0/41.22)	28.18 (0/71.08)	9.05 (0/46.17)	3.03 (0/14.26)
$d_0(x)$	0 (0/81.48)	0 (0/2)	0 (0/0)	0 (0/0)	0 (0/0)
$d_1(x)$	68.87 (0/27.89)	0.40 (0/54.34)	0 (0/0.10)	0 (0/0)	0 (0/0)
$d_2(x)$	82.21 (0/15.81)	17.88 (0/58.11)	0.22 (0/5.08)	0 (0/0.07)	0 (0/0)
$s(x)$	0 (0/59.19)	0 (0/0.33)	0 (0/0)	0 (0/0)	0 (0/0)

Table 4.23: Results for the refined error estimate in Equation 4.27.

despite the pessimistic error estimate and the singularity $s(x)$ did not
return a single false positive for either estimate. What is interesting
about these error estimates is that they never under-estimated the error,
resulting in no false positives at all.

4.3.3   Summary

From the descriptions of the behavior of the individual error estimators,
we can extract some general observations:

1. The higher the degree of the quadrature rules used to compute a
   linear error estimate, the smaller the probability of an "accidentally
   small" error estimate (*e.g.* Gander and Gautschi's vs. Piessens *et
   al.* 's error estimators),

2. The higher the number of nodes by which two quadrature rules
   differ, the more sensitive the error estimate computed from their
   difference reacts to non-interpolable integrands (*e.g.* Berntsen and
   Espelid's 1984 error estimator vs. Piessens *et al.* 's),

3. Based on the two first observations, doubly-adaptive quadrature
   algorithms which rely on initial error estimates from lower-degree
   rules are only as reliable as the lowest-degree error estimate (*e.g.* Oliver's,
   de Boor's and Venter and Laurie's error estimators),

4. Error estimators using non-linear extrapolation usually rely more
   than the linear error estimates on assumptions which can not be verified, which are valid for *most* but not *all* (*e.g.* discontinuous or singular) integrands (*e.g.* Laurie's 1983 error estimator and de Boor's
   CADRE).

According to the results using the chosen test integrands, the best two
error estimators appear to be that of Piessens *et al.* (Section 2.17) which
is the error estimator for the adaptive routines in the popular integration
library QUADPACK, and the two new error estimators presented herein
(Section 4.2).

The relatively few false positives returned by the QUADPACK error estimator may seem negligible in contrast with its efficiency (evidenced by

the much smaller percentage of false negatives) compared to the new error estimate. We can verify this by evaluating the smooth integral

$$\int_1^2 \frac{0.1}{0.01 + (x - \lambda)^2} \, dx$$

first suggested by Lyness in [63], for which compute 1 000 realizations of the parameter $\lambda \in [1, 2]$. We use both Piessens *et al.* 's error estimate with 10 and 21 point Gauss and Gauss-Kronrod rules respectively and the new error estimates using both a pair of 11 and 21-point Clenshaw-Curtis quadrature rules for the trivial error estimate and an 11-point Clenshaw-Curtis quadrature rule with $\vartheta_1 = 1.1$ for the more refined estimate, all within a recursive scheme such as in Algorithm 1 with $\tau' = \tau/\sqrt{2}$, to a relative precision of $\tau = 10^{-9}$. On average, Piessens *et al.* 's error estimate requires 157 function evaluations while the new error estimates require 379 and 330 evaluations respectively – roughly twice as many. Both methods integrate all realizations to the required tolerance.

If we consider, however, the Waldvogel[5] function

$$W(x) = \int_0^x \lfloor e^t \rfloor \, dt$$

which we wish to evaluate to the relative precision $\tau = 10^{-9}$ for 1 000 realizations of $x \in [2.5, 3.5]$ using both the error estimates of Piessens *et al.* and our new error estimators as described above, we get very different results. While Piessens *et al.* 's error estimator fails in roughly three quarters of all cases (753 failures out of 1 000, see Fig. 4.9), usually missing a sub-interval containing one or more discontinuities and using, on average, 29 930 function evaluations, our new error estimators succeeds on every trial, using on average 31 439 and 29 529 function evaluations respectively. The cautious estimate, in this case, pays off.

4.4 Analysis

The reason in this increased reliability is best explained by considering, for any error estimator, the set of integrands for which it will *always* fail.

---

[5]This function was suggested to the author by Prof. Jörg Waldvogel.

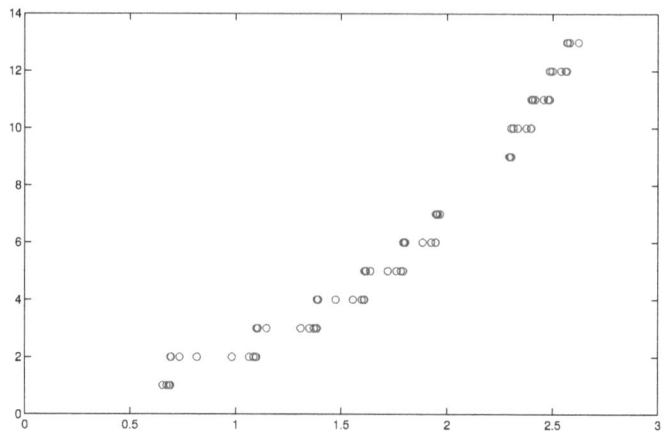

Figure 4.9: Piessens *et al.* 's error estimate used to evaluate one realization of the Waldvogel-function (green). The blue circles mark the edges of the sub-intervals. Note that the integrand is not well resolved near ∼ 2.1.

Consider the polynomials orthogonal with respect to the discrete product

$$(p_i(x), p_j(x)) = \sum_{i=0}^{n} p_i(x_i)p_j(x_j), \quad 0 \le i, j < n \qquad (4.36)$$

where the x_i are the nodes of the quadrature rule or the combined nodes of all the quadrature rules used in the computation of the error estimate in the interval. In the following, when we refer to a pair of functions being orthogonal, we understand them to be orthogonal with respect to the above product.

For any *linear* error estimate relying on the difference between two quadrature rules over the nodes x_i, the error estimate can be computed as

$$\varepsilon = \sum_{i=1}^{n} \eta_i f(x_i)$$

where the η_i are the difference of the weights of the two quadrature rules used in the error estimate for each node[6]. Let $\eta(x)$ be the polynomial

---

[6]The η_i are, incidentally, the weights of a null rule, such as they are constructed by Lyness in [60].

interpolating the η_i at the nodes x_i, $i = 1 \ldots n$. The error can then be computed as the product in Equation 4.36 applied to the integrand $f(x)$ and the polynomial $\eta(x)$:

$$\varepsilon = (\eta(x), f(x)) .$$

Therefore, if the integrand $f(x)$ is of algebraic degree *higher* than that of the quadrature rule used — and will therefore not be correctly integrated — and the integrand $f(x)$ is *orthogonal* to the polynomial $\eta(x)$, then the linear error estimate will be zero and therefore it will *fail*.

For the error estimate of O'Hara and Smith (Section 2.8), which uses more than one derivative, the error estimate fails when the integrand $f(x)$ is of higher algebraic degree than the basic quadrature rule and the coefficients $\tilde{c}_n$, $\tilde{c}_{n-2}$ and $\tilde{c}_{n-4}$ are zero (see Equation 2.33). This is the case when the integrand $f(x)$ is orthogonal to the Chebyshev polynomials $T_n(x)$, $T_{n-2}(x)$ and $T_{n-4}(x)$.

For the error estimate of Berntsen and Espelid (Section 2.19), the error estimate fails when the integrand $f(x)$ is of higher algebraic degree than the basic quadrature rule and the integrand $f(x)$ is orthogonal to the last $2(K - 1)$ null rules[7].

For the non-linear error estimates discussed in Section 4.1.2, the error estimates will fail under similar circumstances: In de Boor's CADRE (see Section 2.9), it is sufficient that the difference between two neighboring entries in the T-table is zero for the error estimate to fail. For a T-table of depth ℓ, this engenders $\mathcal{O}(\ell^2/2)$ different polynomials to which the integrand *may* be orthogonal to for the error estimate to fail.

In the case of Rowland and Varol's or Venter and Laurie's error estimates (see Section 2.10), a difference of zero between two consecutive pairs of rules is sufficient for the error estimate to fail and thus, as for the simple error estimators discussed above, for a sequence of m rules, there are $m - 1$ polynomials to which an integrand $f(x)$ *may* be orthogonal to for which the error estimator will always fail.

In Laurie's error estimate (see Section 2.16), either $Q_\alpha^{(2)} - Q_\beta^{(2)}$ or $Q_\alpha^{(2)} -$

---

[7]In Berntsen and Espelid's original error estimate 2 null-rules are used to compute each E_k (see Equation 2.105) from which the K ratios r_k (see Equation 2.106) are computed. It is, however, only necessary that the nominators of the ratios be zero, hence only $2(K - 1)$ null-rules need to be zero for the estimate to be zero.

$Q_\alpha^{(1)}$ need to be zero for the estimate to fail, resulting in two polynomials to which the integrand *may* be orthogonal to for the error estimate to fail. Similarly, for Favati *et al.*'s error estimate (see Section 2.16), there are three such polynomials.

Finally, for de Doncker's error estimate (see Section 2.14), the case is somewhat more complicated due to the global approach of the algorithm. Since it uses, locally, Piessens *et al.*'s local error estimate (see Section 2.17), it will fail whenever this estimate fails, making it vulnerable to the same family of integrands. Additionally, it will fail whenever the difference between two *global* estimates $\hat{Q}_n^{(m)}[a, b] - \hat{Q}^{(m-1)}[a, b]$ accidentally becomes zero, causing the algorithm to fail *globally*.

For both new error estimates presented here (Equation 4.20 and Equation 4.27), the matter is a bit more complicated. Given two interpolations $g_{n_1-1}^{(1)}(x)$ and $g_{n_2-1}^{(2)}(x)$, with $n_2 \geq n_1$, over the nodes $x_i^{(1)}$, $i = 1 \ldots n_1 - 1$ and $x_i^{(2)}$, $i = 1 \ldots n_2 - 1$ respectively, we define the joint set of n_u nodes $x^{(u)} = x^{(1)} \cup x^{(2)}$ which we will use for the product in Equation 4.36. Given the inverse Vandermonde-like matrices $\mathbf{U}^{(1)} = (P^{(1)})^{-1}$ and $\mathbf{U}^{(2)} = (P^{(2)})^{-1}$ of size $n_1 \times n_1$ and $n_2 \times n_2$ used to compute the coefficients of $g_{n_1}^{(1)}(x)$ and $g_{n_2}^{(2)}(x)$, we can stretch them to size $n_2 \times n_u$ such that

$$\mathbf{c}^{(1)} = \tilde{\mathbf{U}}^{(1)}\mathbf{f}^{(u)}, \quad \mathbf{c}^{(2)} = \tilde{\mathbf{U}}^{(2)}\mathbf{f}^{(u)}$$

where $\tilde{\mathbf{U}}^{(1)}$ and $\tilde{\mathbf{U}}^{(2)}$ are the stretched matrices and $\mathbf{f}^{(u)}$ contains the integrand evaluated at the joint set of nodes $x^{(u)}$. For the error estimate $\|\mathbf{c}^{(1)} - \mathbf{c}^{(2)}\|$ to be zero, $\mathbf{f}^{(u)}$ must lie in the null-space of the $n_2 \times n_u$ matrix

$$\mathbf{U}^{(u)} = \left[\tilde{\mathbf{I}}^{(1)} - \tilde{\mathbf{U}}^{(2)}\right]$$

which has rank r_u equal to the smaller of the number of nodes *not* shared by both $x^{(1)}$ and $x^{(2)}$, i.e. $x^{(u)}\setminus\{x^{(1)} \cap x^{(2)}\}$ or n_2. For the error estimate to be zero, the product $(\mathbf{P}^{(n)})^{-1}\mathbf{f}^{(u)}$ must be zero. This is the case when the integrand $f(x)$ is of algebraic degree $> n_2$ and orthogonal to the r_u polynomials generated by interpolating the values of the first r_u rows of $\mathbf{U}^{(u)}$ at the nodes $x^{(u)}$. If, additionally, the integrand is of degree $> n_2$, then both error estimates will fail.

The space of functions that will cause any of the error estimators presented here to fail is, in essence, infinite, yet for each type of error estimator, this infinite space is subject to different *restrictions*. For the simple **linear** error estimators which compute a *single* divided difference, the space is restricted by a *single* orthogonality restriction. In the case of error estimators such as O'Hara and Smith's or Berntsen and Espelid's, the space is restricted by *three or four*[8] orthogonality restrictions. Instead of being subject to *one or more* restrictions, the space of functions that will cause the **non-linear** error estimators discussed in Section 4.1.2 to fail is *larger* than that of the simple error estimators, since the integrand needs only to be orthogonal to *any* of a set of polynomials for the algorithm to fail. The set of functions for which they will fail is therefore the *union* of a set of functions, each subject to only *one* restriction. For our **new** error estimators, the number of restrictions depends on the number of nodes used. For the trivial error estimate (Equation 4.20), if the nodes $x^{(1)} \subset x^{(2)}$ and $n_2 \approx 2n_1$ (*i.e.* if Clenshaw-Curtis or Gauss-Kronrod rule pairs are used), the number of restrictions will be $\approx n_2/2$. For the more refined error estimate (Equation 4.27), if the basic rule does not re-use more than $\lceil n/2 \rceil$ of its n nodes in each sub-interval, the number of restrictions will be at least $n - 1$.

The new error estimates presented in Section 4.2 are therefore more reliable since the space of functions for which it will fail, albeit infinite, is *more restricted* than that of the other error estimators presented here. It is also interesting to note that if we were to *increase the degree* of the underlying quadrature rules in all our error estimates, the number of restrictions to the space of functions for which they will fail *would not grow*, whereas for our new error estimates, the number of restrictions *grows linearly* with the degree of the underlying quadrature rule.

---

[8]In Berntsen and Espelid's original error estimate, a constant $K = 3$ is used.

Chapter 5

Some Common Difficulties

5.1 Singularities and Non-Numerical Function Values

Since most adaptive quadrature algorithms are designed for general-purpose use, they will often be confronted with non-integrable integrands or integrands containing singularities or undefined values.

Singularities in the integrand can cause problems on two levels:

- The quadrature rule has to deal with a non-numerical value such as a NaN or ±Inf,

- The integrand may not be as smooth and continuous as the algorithm might assume.

Such problems arise when integrating functions such as

$$\int_0^h x^\alpha \, \mathrm{d}x, \quad \alpha < 0$$

which have a singularity at $x = 0$, or when computing seemingly innocu-
ous integrals such as

$$\int_0^h \frac{\sin x}{x} \, dx$$

for which the integrand is undefined at $x = 0$, yet has a well-defined limit

$$\lim_{x \to 0} \frac{\sin x}{x} = 1.$$

In both cases problems *could* be avoided by either shifting the integration
domain slightly or by modifying the integrand such as to catch the undefined cases and return a correct numerical result. This would, however,
require some prior reflection and intervention by the user, which would
defeat the purpose of a general-purpose quadrature algorithm.

Most algorithms deal with singularities by ignoring them, setting the
offending value of the integrand to 0 [13, Section 2.12.7]. Another approach, taken by quad and quadl in Matlab, is to shift the edges of the
domain by $\varepsilon_{\text{mach}}$ if a non-numerical value is encountered there and to
abort with a warning if a non-numerical value is encountered elsewhere
in the interval. Since singularities may exist explicitly at the boundaries
(*e.g.* integration of x^α, $\alpha < 0$ in the range $[0, h]$), the explicit treatment
of the boundaries is needed, whereas for arbitrary singularities within the
interval, the probability of hitting them exactly is somewhat small.

QUADPACK's QAG and QAGS algorithms take a similar approach: since
the nodes of the Gauss and Gauss-Lobatto quadrature rules used therein
do not include the interval boundaries, non-numerical values at the interval boundaries will be implicitly avoided. If the algorithms has the
misfortune of encountering such a value inside the interval, it aborts.

Our approach to treating singularities will be somewhat different: instead of setting non-numerical values to 0, we will simply *remove* that
node from our interpolation of the integrand. This can be done rather
efficiently by computing the interpolation as shown before using a function value of $f(x_j) = 0$ for the offending jth node and then *down-dating*
(as opposed to up-dating) the interpolation, *i.e.* removing the jth node
from the interpolation, resulting in an interpolation of degree $n - 1$:

$$g_{n-1}(x) = \sum_{i=0}^{n-1} c_i^{(n-1)} p_i(x)$$

which still interpolates the integrand at the remaining n nodes. The coefficients $c_i^{(n-1)}$ of $g_{n-1}(x)$ can be computed, as described in Section 3.2.1, using

$$c_i^{(n-1)} = c_i - \frac{c_n}{b_n^{(n-1)}} b_i^{(n-1)}, \quad i = 0 \dots n$$

where the c_i are the computed coefficients of $g_n(x)$. The $b_i^{(n-1)}$ are the coefficients of the downdated Newton-polynomial computed by solving the upper-triangular system of equations

$$
\begin{pmatrix}
\alpha_0 & -(x_j+\beta_1) & \gamma_2 & & \\
& \ddots & \ddots & \ddots & \\
& & \alpha_{n-2} & -(x_j+\beta_{n-1}) & \gamma_n \\
& & & \alpha_{n-1} & -(x_j+\beta_n) \\
& & & & \alpha_n
\end{pmatrix}
\begin{pmatrix}
b_0^{(n-1)} \\
b_1^{(n-1)} \\
\vdots \\
b_{n-1}^{(n-1)}
\end{pmatrix}
=
\begin{pmatrix}
b_1 \\
b_2 \\
\vdots \\
b_n
\end{pmatrix}
$$

using back-substitution, where the b_i are the coefficients of the Newton polynomial over the nodes of the quadrature rule (Equation 4.23). α_i, β_i and γ_i are the coefficients of the three-term recurrence relation satisfied by the polynomials of the orthogonal basis:

$$\alpha_k p_{k+1}(x) = (x + \beta_k)p_k(x) - \gamma_k p_{k-1}(x).$$

For the case of the normalized Legendre polynomials used here, the coefficients are

$$\alpha_k = \sqrt{\frac{(k+1)^2}{(2k+1)(2k+3)}}, \quad \beta_k = 0, \quad \gamma_0 = \sqrt{2}, \quad \gamma_k = \sqrt{\frac{k^2}{4k^2-1}}$$

The modified vectors $\mathbf{c}^{(n-1)}$ and $\mathbf{b}^{(n-1)}$ are then used in the same way as $\mathbf{c}$ and $\mathbf{b}$ respectively for the computation of the integral and of the error estimate.

5.2 Divergent Integrals

In the following, we will refer to *divergent integrals* as integrals which tend to $\pm\infty$ and thus cause most algorithms to either recurse infinitely or return an incorrect finite result. Divergent integrals are usually caught by limiting the recursion depth or the number of function evaluations artificially. Both approaches do not *per se* attempt to detect divergent behavior, and may therefore cause the algorithm to fail for complicated yet non-divergent integrals.

In [73], Ninham studied the approximation error when computing

$$\int_0^h x^\alpha \, \mathrm{d}x \tag{5.1}$$

using the trapezoidal rule. The integral exists for $\alpha > -1$ and is divergent otherwise. Following his analysis, we compute the refined error estimate described in Section 4.2 (Equation 4.29) for the intervals $[0, h]$ and $[0, h/2]$ using an 11-node Clenshaw-Curtis quadrature rule and removing the singular node at $x = 0$ as described above.

We note that as the algorithm recurses to the leftmost interval, the local error estimate as well as the computed integral itself remain constant for $\alpha = -1$ and *increase* for $\alpha < -1$. In Figure 5.1 we plot the ratio of the error estimate in the left sub-interval versus the error in the entire interval, $\varepsilon[0, h/2]/\varepsilon[0, h]$, over the parameter α. For $\alpha = -1$ the ratio is 1 meaning that the error estimate of the leftmost interval remains constant even after halving the interval. For $\alpha < -1$, for which the integral diverges, the error estimate in the left half-interval is *larger* than the error estimate over the entire interval.

The rate at which the error decreases (or, in this case, increases) may be a good indicator for the convergence or divergence of the integral in Equation 5.1, where the singularity is at the boundary of the domain, yet it does not work as well for the shifted singularity

$$\int_0^h |x - \beta|^\alpha \, \mathrm{d}x, \quad \beta \in [0, h/2]. \tag{5.2}$$

Depending on the location of the singularity ($x = \beta$), the ratio of the error estimates over $[0, h]$ and $[0, h/2]$ varies widely for both $\alpha > -1$ and $\alpha \le -1$ and can not be used to determine whether the integral diverges or not (see Figure 5.2).

In Figure 5.2 we have shaded the regions in which the ratio of the error estimates $\varepsilon[0, h/2]/\varepsilon[0, h] > 1$ for different values of α and the location of

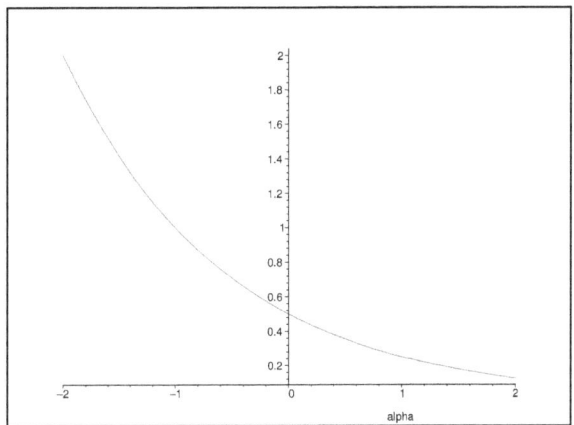

Figure 5.1: Ratio of the error estimates for $\int_0^h x^\alpha \, dx$ for the intervals $[0, h/2]$ over $[0, h]$ for different α. Note that the error grows (ratio > 1) for $\alpha < -1$, where the integral is divergent.

the singularity β. For this ratio to be a good indicator for the value of α (and hence the convergence/divergence of the integral), the shaded area should at least partially cover the lower half of the plot where $\alpha < -1$, which it does not.

A more reliable approach consists of comparing the computed *integral* in two successive intervals $[a, b]$ and $[a, (a + b)/2]$ or $[(a + b)/2, b]$. For the integrand in Equation 5.1, the integral in the left sub-interval $[0, h/2]$ is *larger* than that over the interval $[0, h]$ when $\alpha < -1$. For the integral in Equation 5.2 the ratio of the integrals in the intervals $[0, h]$ and $[0, h/2]$ is larger than 1 for *most* cases where $\alpha \leq -1$ (see Figure 5.3).

Although this relation (ratio $> 1 \Rightarrow \alpha < -1$), which is independent of the interval h, is not always correct, it can still be used as a statistical *hint* for the integrand's behavior during recursion. In the interval $-2 \leq \alpha \leq -1$ it is correct approximately two thirds of the time. We will therefore count the number of times that

$$\frac{Q_n[a, (a + b)/2]}{Q_n[a, b]} \geq 1 \quad \text{or} \quad \frac{Q_n[(a + b)/2, b]}{Q_n[a, b]} \geq 1 \tag{5.3}$$

during recursion. Note that since the integral tends to either $+\infty$ or $-\infty$,

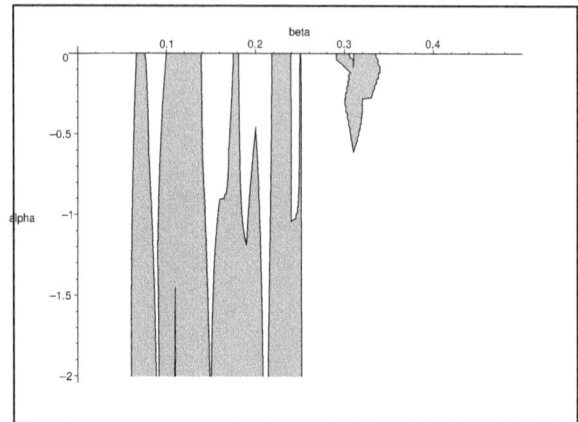

Figure 5.2: Contour of the ratio of the error estimates for $\int_0^h |x - \beta|^\alpha \, \mathrm{d}x$ over the intervals $[0, 1]$ and $[0, 1/2]$. The filled area represent the region in which this ratio is larger than 1.

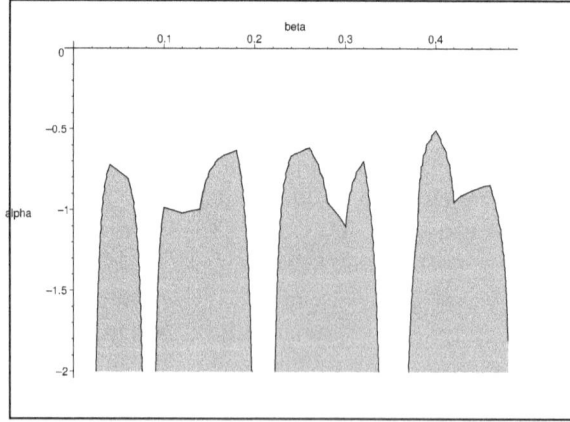

Figure 5.3: Contour of the ratio of the integral estimates for $\int_0^h |x - \beta|^\alpha \, \mathrm{d}x$ over the intervals $[0, 1]$ and $[0, 1/2]$. The filled area represent the region in which this ratio is larger than 1.

the integral approximations will be of the same sign and thus the sign of the ratios do not matter. If this count exceeds some maximum number and is more than half of the recursion depth – *i.e.* the ratio of integrals was larger than one over more than half of the subdivisions – then we declare the integral to be divergent and return an error or warning to the user to this effect.

This approach will not work for all types of divergent integrals. Consider the integral

$$\int_0^b \frac{\log(1/x)^{-1-\alpha}}{x} \, \mathrm{d}x \quad = \quad \frac{\log(1/b)^{-\alpha}}{\alpha}, \quad b < 1$$

which does not exist for $\alpha = 0$. As opposed to singularities of the type in Equation 5.2, however, the ratios in Equation 5.3 are all < 1 although the integral goes to ∞. Such singularities could, in theory, be caught by checking the convergence of the sequence of estimates

$$Q[0,h], \ Q[0,h/2], \ Q[0,h/4], \ \ldots$$

yet for each type of convergence test it would be possible to create families of integrals for which such a test would fail.

Chapter 6

The Quadrature Algorithms

6.1 Design Considerations

Although the main foci of this thesis are function representation (see Chapter 3), error estimation (see Chapter 4) and the treatment of singular or divergent integrals (see Chapter 5), there are a number of implementation details that have to be considered before implementing these new features in an adaptive quadrature algorithm.

In 1975, Rice [85] describes several common components of quadrature algorithms:

- "*Interval Processor Components*": The basic quadrature rule or set of rules and the associated error estimator,

- "*Bound Estimator Components*": Effectively the distribution of the global tolerance over the sub-intervals,

- "*Special Behaviour Components*": Possible detection of noise, discontinuities and singularities,

- "*Interval Collection Management Components*": The order in which intervals are processed,

Arguing that there are several different possibilities for the implementation of each component, he concludes that there are possibly *"1 to 10
million"* potentially interesting combinations thereof worth studying.
Since we want to test the new error estimates and the treatment of non-
numerical values of the integrand and divergent integrals, we will not test
the several million alternatives, but make some common design decision
based both on published results and our own observations.

6.1.1   Recursive vs. Heap Structure

The main difference between the general Algorithms presented in Chapter 1, Algorithm 1 and Algorithm 2, is in the strategy they use to select
the next interval for processing.

Rice calls this process the *"interval collection management component"*
and specifies four types of such components:

- **Stack**: The intervals are generated and traversed *depth-first* from
  left to right (see Figure 6.1). This can be done either implicitly
  using recursion as in Algorithm 1 or explicitly as is done in McK-
  eeman and Tesler's integrator (see Section 2.4).

- **Heap**: The intervals are sorted by decreasing error estimate and
  the interval with the largest error estimate is processed, as is done
  in Algorithm 2.

- **Queue**: A FIFO-queue of all intervals is maintained and processed
  in-order. This is equivalent to a *breadth-first* traversal of the intervals.

- **Boxes**: The intervals are grouped into boxes for different ranges of
  the error estimate. Intervals are selected as with the queue, yet no
  sorted structure is preserved.

The *stack* has the advantage that it is relatively simple to implement
in programming environments where recursion is permitted and that it
does not require that we store data for more intervals than the maximum
recursion depth.

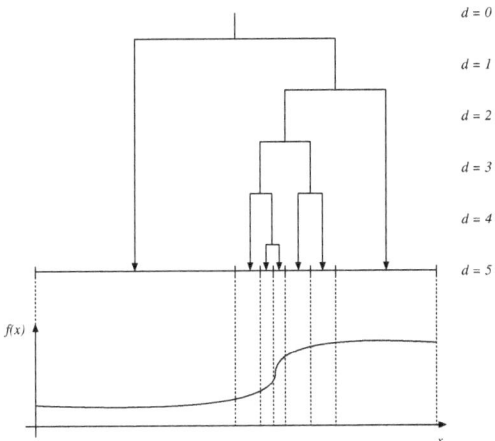

Figure 6.1: The intervals generated during adaptive quadrature.

It also has the disadvantage that during recursion, the algorithm must allocate a portion of the allowed total error to each interval. This is usually done by selecting a local tolerance τ_k such that

$$\tau_k = \tau \frac{b_k - a_k}{b - a} \quad \text{or} \quad \tau_k = \tau 2^{-d/2}.$$

In the former case, we can guarantee that the sum of the local errors will be below the global tolerance. In the latter case we guarantee that, when using bisection, the sum of the local errors will have the same *variance* as the global error (see Section 2.4). Different variations of how the local tolerance can be allocated can be found in O'Hara and Smith's integrator (see Section 2.8), Oliver's integrator (see Section 2.11), Garribba *et al.* 's integrator (see Section 2.13) or Ninomiya's integrator (see Section 2.15).

In Equation 2.8 and Equation 2.9 we saw how this could go wrong, causing the algorithm to fail to converge even for simple integrands. This problem is inherent to stack-based algorithms and can only be alleviated therein by using a less strict (or no) local tolerance such as is done by Patterson (see Section 2.12) or Gander and Gautschi (see Section 2.20). These approaches, however, do not guarantee that the sum of the error estimates will be below the global tolerance.

The *queue*, traverses the tree in Figure 6.1 *breadth-first* and requires only as much memory as the maximum number of intervals at any recursion level. In [92], Shapiro implements an adaptive quadrature based on Forsythe, Malcolm and Moler's QUANC8 (see Section 2.7) using a queue and notes that whereas the stack, queue and heap all generate and traverse the same tree of intervals (see Figure 6.1), and therefore generate the same result, the queue generates better results than the stack when the number of allowed function evaluations is limited, *i.e.* when the algorithm is terminated before reaching the required tolerance, and is more efficient in terms of memory requirements and maintenance than the heap.

The queue, however, has the same problem as the stack in that we must also decide, at each recursion level, which intervals to discard. Although this decision can be made considering *all* the intervals at a given recursion depth, the allocation of error or tolerance is made nonetheless, which may cause the algorithm to fail to converge. Shapiro, for example, discards an interval only when the error estimate is less than the local tolerance

$$\tau_k = \tau \frac{b_k - a_k}{b - a}.$$

This is equivalent to the scaling of the local tolerance used by Kuncir (see Section 2.2) and by McKeeman (see Section 2.3) and thus, as with McKeeman's algorithm, the queue-based algorithm will *also* fail to converge for the left-most interval of the problems in Equation 2.8 and Equation 2.9. This problem arises whenever we have to discard an interval and can therefore only be avoided if we do not discard *any* intervals. This is essentially what both the *heap* and the *boxes* do. The only difference between the two is that the boxes sort the intervals with a larger granularity.

The main drawback of such an approach is of course that the heap or the boxes grow with every interval subdivision. The resulting memory requirements and the overhead to manage the heap or boxes make these approaches somewhat less attractive than the stack or queue. The first algorithm using a heap only appeared in 1973 (see Section 2.12), when computer memories were less restricted, making such an approach practicable.

The use of a heap was shown, in a direct comparison by Malcolm and Simpson [66], to be more efficient than the use of a stack since the error in intervals that may not have reached the local tolerance requirements used by the stack can be compensated for by intervals who's error estimates are well below the local requirements, thus requiring no further refinement. For the algorithms presented herein, we will use, in principle, a heap to store the intervals and select the interval with the largest error estimate for further processing. To avoid the heap growing indefinitely, we will remove the following intervals:

- Intervals who's width no longer contains sufficient machine numbers such that the nodes of the quadrature are distinct,

- Intervals who's error estimate is smaller than the precision of the computation of the integral,

- The interval with the smallest error estimate if the heap exceeds a pre-determined size.

The first condition is detected when sub-dividing an integral by verifying that

$$\tilde{x}_0 < \tilde{x}_1 \quad \wedge \quad \tilde{x}_{n-1} < \tilde{x}_n, \quad \tilde{x}_i = \frac{a+b}{2} + \frac{b-a}{2}x_i, \quad x_i \in [-1, 1]$$

i.e. by verifying that the nodes of the quadrature rule in the integration interval $[a, b]$ are indeed distinct.

The second condition is verified by testing if

$$\varepsilon < |q| \operatorname{cond}(\mathbf{P}^{-1}) \, \varepsilon_{\mathsf{mach}}$$

where $\varepsilon_{\mathsf{mach}}$ is the machine epsilon, ε and q are the computed error estimate and integral respectively and $\mathbf{P}^{-1}$ is the inverse Vandermonde-like matrix used to compute the coefficients from which the integral is computed.

The final condition is verified by keeping tabs not only on the interval with the *largest* error estimate, but also on the interval with the *smallest*. If the heap grows beyond a maximum size, the interval with the smallest error is dropped.

In any case, every time an interval is dropped, its approximation to the integral and the integration error are stored in the excess variables q_{xs} and

ε_{xs} respectively. The total integral and error estimate are then computed
as

$$q_{tot} = \sum_{k \in H} q_k + q_{xs}, \quad \varepsilon_{tot} = \sum_{k \in H} \varepsilon_k + \varepsilon_{xs}$$

where $\sum_{k \in H}$ is the summation over all elements of the heap. The algorithms terminate if either $\varepsilon_{tot} < \tau$ or, if $\varepsilon_{xs} > \tau$ and thus the required tolerance can not be achieved, if $\sum_k \varepsilon_k < \tau$.

Since we need to keep tabs of both the intervals with the minimum and maximum error estimate, we only maintain a heap *in principle*. In practice, we maintain an unsorted array of intervals which is searched linearly for the maximum and minimum elements in each iteration. This has the advantage that the sums in Equation 6.1.1 can be computed in every iteration, thus avoiding truncation and cancellation problems that would arise when updating q_{tot} and ε_{tot}.

6.1.2   The Choice of the Basic Rule

Most of the algorithms presented in Chapter 2 differ principally in the type of basic quadrature rule (or rules) used to compute both the integral and the error approximation. The most commonly used types of rules are:

- **Newton-Cotes** quadrature rules with $x_i = -1 + 2i/n$, $i = 0 \dots n$,

- **Recursively Monotone Stable** (RMS) quadrature rules where the distances between the nodes are all integer multiples of the smallest distance and where all the nodes can be re-used after bisection,

- **Clenshaw-Curtis** type quadrature rules where the nodes are either the roots or the extrema of a Chebyshev polynomial in either $[-1, 1]$ or $(-1, 1)$,

- **Gauss** or **Gauss-Lobatto** quadrature rules and their **Kronrod** extensions where the nodes x_i, $i = 0 \dots n$ are the roots of the $n + 1$st orthogonal polynomial relative to some measure.

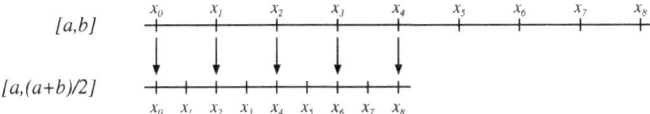

Figure 6.2: Re-use of nodes for a 9-point Newton Cotes rule after bisection.

The **Newton-Cotes rules** have the advantage that the nodes can easily be re-used after doubling n or bisecting or otherwise sub-dividing the interval (see Figure 6.2).

The main disadvantage of Newton-Cotes rules is that they are numerically unstable: the 9-point Newton-Cotes quadrature rule and *all* rules with more than 11 nodes contain one or more negative weights, which can lead to cancellation when evaluating the quadrature rule. As the number of nodes n grows, the condition number of the rule, as defined by Engels [17], grows to ∞:

$$\lim_{n \to \infty} \sum_{i=0}^{n} |w_i^{(n)}| = \infty.$$

Newton-Cotes rules are therefore only used in adaptive schemes for small n.

A compromise between the re-usability of the nodes and the condition number of the resulting rule is achieved with the **RMS rules** described by Favati *et al.* [26]. All the nodes of each RMS rule can be re-used after bisection (see Figure 6.3) or when increasing the multiplicity or degree of a rule within the same interval *and* the weights of the resulting quadrature rules are always positive and hence the rules are numerically stable.

Favati *et al.* identify 74 such rules for several odd n up to $n = 249$. The rules can not be constructed on the fly and are generated by an exhaustive search and tabulated for latter use.

The **Clenshaw-Curtis type rules** were first introduced by Fejér in 1933 [28] and popularized by Clenshaw and Curtis in 1960 [10]. These rules

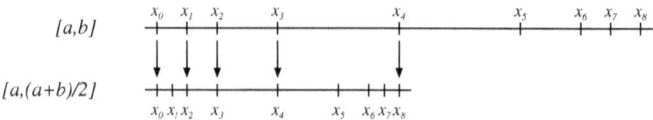

Figure 6.3: Re-use of nodes for a 9-point RMS rule after bisection.

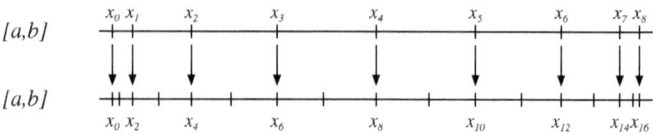

Figure 6.4: Re-use of nodes for a 9-point Clenshaw-Curtis rule in the 17-point Clenshaw-Curtis rule.

exist in three variants

$$\text{Clenshaw-Curtis rule: } x_i = -\cos\left(\frac{i\pi}{n}\right)$$

$$\text{Fejér's "first rule": } x_i = -\cos\left(\frac{(i+1/2)\pi}{n+1}\right)$$

$$\text{Fejér's "second rule": } x_i = -\cos\left(\frac{(i+1)\pi}{n+2}\right)$$

for $i = 0\ldots n$. The quadrature rules, which are studied extensively in [75], all have positive weights and are numerically stable.

The weights of the Clenshaw-Curtis rules can be constructed efficiently using the Fast Fourier Transform [40, 101]. Although the nodes can generally not be re-used after bisection[1], the nodes can be re-used if the rule over $2n + 1$ points in the same interval is also computed (see Figure 6.4).

Finally, **Gauss quadrature rules**, which are used in most quadrature applications today, offer a degree of $2n + 1$ while requiring only $n + 1$ nodes. Although the nodes of the rules can not be re-used after bisection and are not part of any higher-degree Gauss quadrature rule, the rules can

---

[1] exceptions include the edge nodes in the Clenshaw-Curtis rule for even n or the central nodes when n is a multiple of 6.

be extended using Kronrod extensions [54], in which a Gauss quadrature rule over $n+1$ points is extended by $n+2$ points resulting in a quadrature rule of degree $3n + 2$. The weights of the Gauss quadrature rules and those of their Kronrod extensions are always positive and thus the rules are numerically stable [39, Theorem 1.46].

The advantage in the degree of precision of Gauss quadrature rules over Clenshaw-Curtis quadrature rules is, however, as is shown by Trefethen in [95], does not translate into a corresponding advantage in the rate of convergence for functions which are not analytic in a large neighborhood around the interval of integration.

Since, in the algorithms presented here, we will not just evaluate a quadrature rule, but compute an interpolation, we are much more interested in the *interpolatory* characteristics of the nodes. In this vein, the nodes of Fejér's first rule are most interesting since given the interpolation error

$$f(x) - g_n(x) = \frac{f^{(n+1)}(\xi)}{(n+1)!} \prod_{i=0}^{n}(x - x_i), \quad \xi \in [a, b] \qquad (6.1)$$

for $f(x)$ $n+1$ times continuously differentiable in $[a, b]$, the nodes, which are the roots of the nth Chebyshev polynomial, *minimize the maximum* of the monic polynomial given by the product on the left hand side:

$$\prod_{i=0}^{n}(x - x_i) = \frac{1}{2^n}T_{n+1}(x).$$

The nodes of Fejér's first (and second) rule have one major disadvantage: the rules are "open" and thus the domain boundaries ± 1 are not included in the quadrature rule. As seen in Section 4.3, certain features of the integrand near the boundaries could thus be missed. This is, of course, an advantage if we can not deal with singularities or non-numerical function values which often appear at the interval edges, but, as we will see later, we will deal with such problems differently.

For the nodes of the Clenshaw-Curtis quadrature rule, the polynomial in the interpolation error Equation 6.1 is no longer minimal, but can be represented as

$$C_n(x) = \prod_{i=0}^{n}\left(x - \cos\left(\frac{i\pi}{n}\right)\right) = \frac{1}{2^n}\left(T_{n+1}(x) - T_{n-1}(x)\right)$$

where $T_k(x)$ is the kth Chebyshev polynomial of the first kind[2]. Since for all Chebyshev polynomials, $T_k(x) \in [-1, 1]$ for all $k > 0$ and $x \in [-1, 1]$, this monic polynomial is *at most* twice as large as the one resulting from the nodes of Fejér's first rule

$$\prod_{i=0}^{n} \left(x - \cos\left(\frac{(i + 1/2)\pi}{n + 1} \right) \right) = \frac{1}{2^n} T_{n+1}(x).$$

The interpolation error over the nodes of the Clenshaw-Curtis rule therefore converges at the same *rate* as the error over the (optimal) nodes of Fejér's first rule.

For the nodes of the Gauss quadrature rule, the polynomial on the right hand side of Equation 6.1 is the monic $n + 1$st Legendre polynomial. Since the Legendre polynomials of degree $n \geq 1$ have a maximum of 1 at $x = 1$ and a leading coefficient $k_n = (2n!)/(2^n n!^2)$ [39, p. 27], the monic Legendre polynomials of degree $n \geq 1$ have a maximum value of $(2^n n!^2)/(2n!)$ at $x = 1$, which decays at a *slower* rate than do the maxima of the monic polynomials over the Fejér or Clenshaw-Curtis rules (see Figure 6.5).

---

[2]This can easily be shown by considering that $U_{n-1}(x)$, the Chebyshev polynomial of the second kind, has its roots at the nodes of Fejér's second rule. This rule can be extended to a Clenshaw-Curtis rule by adding the roots at $x = \pm 1$. The polynomial with the added roots can be written as

$$C_n(x) = (x + 1)(x - 1)U_{n-1}(x) = x^2 U_{n-1}(x) - U_{n-1}(x).$$

Using the relations

$$
\begin{aligned}
T_{n+1}(x) &= 2x T_n(x) - T_{n-1}(x) \\
2T_{n+1}(x) &= U_{n+1}(x) - U_{n-1}(x) \\
x U_n(x) &= U_{n+1}(x) - T_{n+1}(x)
\end{aligned}
$$

we can re-write $C_n(x)$ as

$$
\begin{aligned}
C_n(x) &= x(U_n(x) - T_n(x)) - U_{n-1}(x) \\
&= U_{n+1}(x) - U_{n-1}(x) - T_{n+1}(x) - x T_n(x) \\
&= T_{n+1}(x) - \frac{1}{2}(T_{n+1}(x) + T_{n-1}(x)) \\
&= \frac{1}{2}(T_{n+1}(x) - T_{n-1}(x)).
\end{aligned}
$$

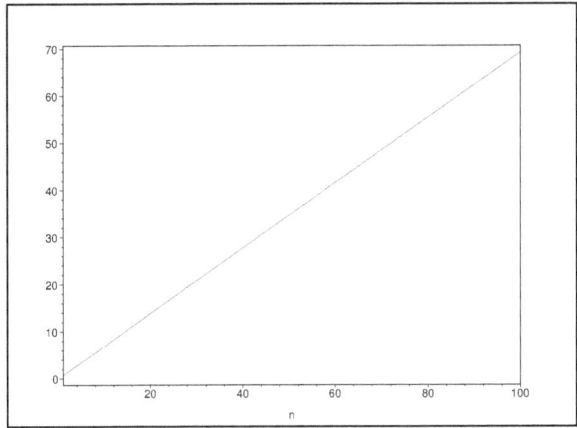

Figure 6.5: $-\ln$ of the maximum value of the nth monic Chebyshev (red) and Legendre (green) polynomials.

Finally, the interpolatory qualities of the equidistant nodes of the Newton-Cotes quadrature rules are undisputedly sub-optimal, as first noted by Runge in 1901 [89]. Henrici also shows in [44, Chapter 5.5] that the condition of the interpolation becomes exceedingly bad for large n (*i.e.* small changes in the data may cause large changes in the interpolant) and Turetskii [97] (referenced in [96]) shows that the Lebesgue constant for an interpolation over the nodes of an $n+1$ point Newton-Cotes rule grows exponentially as

$$\Lambda_n \sim \frac{2^{n+1}}{n \log n}$$

for $n \to \infty$, *e.g.* that the norm of the difference between the constructed interpolation and the optimal interpolation grows exponentially with increasing n.

For the nodes of Clenshaw-Curtis type rules and the Gauss quadrature rules, the resulting Vandermonde-like matrix Equation 3.17 is well-conditioned even for large n (see Figure 6.6). Therefore, for such cases, we do not need to apply any of the special algorithms described in Section 3.2 to compute the coordinates since, for a fixed set of nodes, we can pre-compute the inverse of the Vandermonde-like matrix and compute

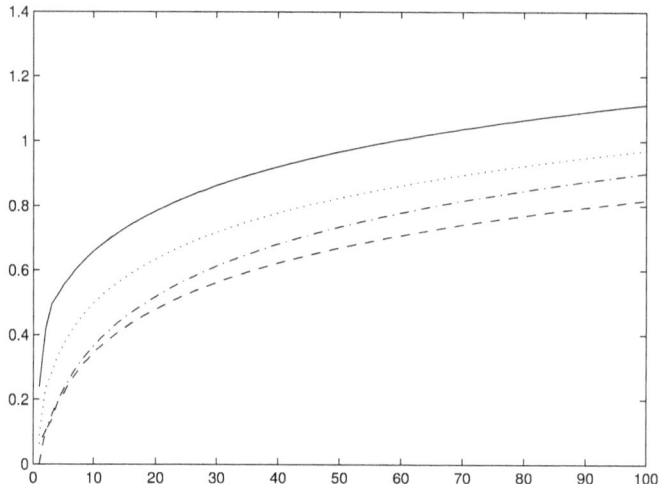

Figure 6.6: $\log_{10}$ of the condition number with respect to the 2-norm of the $(n+1) \times (n+1)$ Vandermonde-like matrix for the orthonormal Legendre polynomials over the nodes of the Clenshaw-Curtis rule (—), Fejér's first rule ($\cdots$), Fejér's second rule ($-\cdot-$) and the Gauss-Legendre quadrature rule ($--$).

the coefficients, using the notation in Section 3.2, with

$$\mathbf{c} = \mathbf{P}^{-1}\mathbf{f}.$$

6.1.3 Handling Non-Numerical Function Values

Since most adaptive quadrature algorithms are designed for general-purpose use, they will often be confronted with integrands containing singularities or undefined values. As we have seen in Section 5.1, these function values can be removed from our interpolation of the integrand by *down-dating* the interpolation.

This is done as is shown in Algorithm 7. If, when evaluating the integrand, a non-numerical value is encountered, the function values is set to zero and the index of the node is stored in nans (Line 4). The coefficients of the interpolation are then computed for those function values (Line 6). For each index in nans, first the coefficients $\mathbf{b}$ of the Newton polynomial over the nodes of the quadrature rule are down-dated as per Equation 3.35 (Line 8) where the upper-triangular matrix U_i^{-1} does not need to be inverted explicitly (see Algorithm 5). The downdated $\mathbf{b}$ is then in turn used to downdate the interpolation coefficients $\mathbf{c}$ as per Equation 3.36 (Line 9).

Algorithm 7 Interpolation downdate procedure

1: nans ← {}   (*initialize* nans)
2: **for** $i = 0 \dots n$ **do**
3:     $f_i \leftarrow f\left((a+b)/2 + x_i(b-a)/2\right)$ (*evaluate the integrand at the nodes*)
4:     **if** $f_i \in \{\mathsf{NaN}, \mathsf{Inf}\}$ **then** $f_i \leftarrow 0$, nans ← nans $\cup$ $\{i\}$ **end if**
       (*if the result is non-numerical, set the node to zero and remember it*)
5: **end for**
6: $\mathbf{c} \leftarrow \mathbf{P}^{-1}\mathbf{f}$ (*compute the initial interpolation coefficients*)
7: **for** $i \in$ nans **do**
8:     $\mathbf{b} \leftarrow \mathbf{U}_i^{-1}\mathbf{b}$ (*downdate the coefficients of the Newton polynomial*)
9:     $\mathbf{c} \leftarrow \mathbf{c} - \frac{c_n}{b_n}\mathbf{b}$ (*downdate the coefficients of the interpolation*)
10:    $n \leftarrow n - 1$ (*decrement the degree*)
11: **end for**

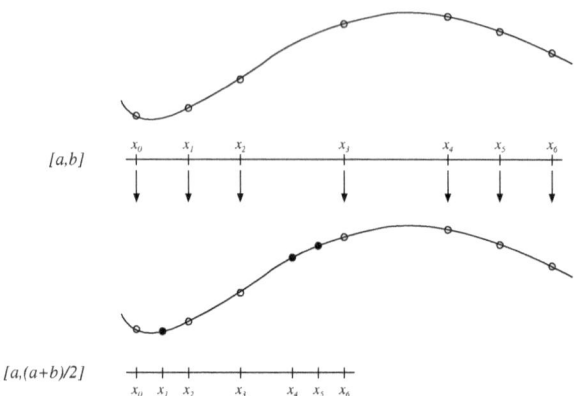

Figure 6.7: An interpolation is computed over the nodes of the quadrature rule in the interval $[a, b]$. After bisection, this interpolation is extended by the function values of the new nodes of the same rule in the sub-interval (filled dots).

6.2 An Adaptive Quadrature Algorithm using Interpolation Updates

The main idea behind this quadrature algorithm is that instead of applying a new quadrature rule in each interval, and thus ignoring information from previous intervals, we could simply *extend* the interpolation computed in the previous interval by *updating* the interpolation with new points inside the sub-interval, as shown in Figure 6.7.

For this algorithm we will use the 9-point RMS rule as described in [26] since for the $n + 1$ point rule we only need to evaluate an additional $n/2$ function values in each sub-interval. The interpolatory qualities of these nodes, while not optimal, are sufficiently good, since the polynomial in the right hand side of the interpolation error Equation 6.1 is at most ~ 4 times that of the optimal set of nodes. Finally, the condition number for the Vandermonde-like matrix for the orthonormal Legendre polynomials at the 9 RMS-nodes is only 8.46.

Given the coefficients $\mathbf{c}^{\text{old}}$ and $\mathbf{b}^{\text{old}}$ from the previous level of recursion as computed by Equation 4.30 (see Section 4.2), we can use Algorithm 4

to update the interpolation using the function values at the new nodes
inside the sub-interval, resulting in the new coefficients $\mathbf{c}$ and $\mathbf{b}$ of degree
$n/2$ higher than the original interpolation. Given these two approxi-
mations of different degree, we can compute the naive error estimate in
Equation 4.20.

The interpolation update and the error estimate only make sense if the
integrand is sufficiently smooth inside the entire interval from which the
interpolation nodes are taken. This can not be tested in practice, but we
can test that

$$\frac{\|\mathbf{c} - \mathbf{c}^{\text{old}}\|}{\|\mathbf{c}\|} < \text{hint} \qquad (6.2)$$

i.e. that the change in the interpolation coefficients is relatively small
when the degree is increased. This is analogous to the hint used by Venter
and Laurie in [99]. If Equation 6.2 does *not* hold, we assume that the
integrand is not sufficiently smooth and restart the interpolation using
only the nodes and function values inside the current interval.

To avoid problems with loss of precision when updating the interpolation,
we will also only allow the polynomial to grow up to a maximum degree
n_{max}. If the degree exceeds n_{max}, we restart the interpolation using only
the nodes and function values inside the current interval.

The entire algorithm is shown in Algorithm 8. In Line 1 the integrand is
evaluated for the initial RMS rule and in Line 2 the initial interpolation
coefficients are computed. The heap H is initialized in Line 5 with this
data and $\mathbf{b}_{\text{def}}$, the pre-computed coefficients of the Newton polynomial
over the nodes of the quadrature rule. While the sum of the error estimates in the heap is larger than the required tolerance τ (Line 7), the
interval with the largest error estimate is popped from the heap (Lines 8
and 9). In each sub-interval, the coefficients of both the interpolation
(Line 12) and the Newton polynomial (Line 13) are transformed to the
sub-interval as described in Equation 3.3. The coefficients are then updated using the new nodes in the sub-interval as described in Section 3.2.1
(Lines 15 to 24), skipping non-numerical function values when encountered (Line 17). If the degree of the resulting interpolation is too high
or the change in coefficients was too large, suggesting that the integrand
is not sufficiently smooth, the interpolation is re-computed only on the
nodes within the sub-interval (Lines 25 to 27). Note that we don't simply
evaluate $\mathbf{c} = \mathbf{P}^{-1}\mathbf{f}$, but instead apply Algorithm 7. The estimates for the

integral (Line 28) and the integration error (Line 29) are then computed
and the algorithm tests for divergence of the integral (Lines 30 and 31).
In Line 33, the new interval is returned to the heap unless its error is
too small or the interval can no longer be sub-divided, in which case its
estimates for the integral and error are stored in the excess variables q_{xs}
and ε_{xs} and the interval is discarded (Line 35).

6.3   A Doubly-Adaptive Quadrature Algorithm using the Naive Error Estimate

In this algorithm we will use the naive error estimate described in Section 4.2 (Equation 4.20) within a doubly adaptive scheme. Although in
Chapter 4 we observed that doubly-adaptive quadrature algorithms are
only as good as the initial error estimates for the lowest-degree rule, we
trust that our error estimate is sufficiently robust and will not cause the
entire algorithm to fail.

In a scheme similar to that used in [99], we will compute interpolations
over $n + 1$ and $2n + 1$ Clenshaw-Curtis nodes and use them to compute
the naive error estimate in Equation 4.20. The interval is sub-divided, as
in Algorithm 8, if the relative difference between the two approximations
is larger than some value hint (see Equation 6.2) or if the degree n has
reached some maximum value. Otherwise, n is doubled.

We use $d_{max} + 1$ rules of degree $n_i = 2n_{i-1}$ starting at some initial degree n_0 The interpolations are computed using the pre-computed inverse
Vandermonde-like matrices $(\mathbf{P}^{(i)})^{-1}$, $i = 0 \ldots d_{max}$.

The algorithm is shown in Algorithm 9. In Lines 1 to 4, the coefficients
of the two highest-degree rules are computed and used to approximate
the initial integral and error estimate. In Line 5 the heap H is initialized
with this interval data. The algorithm then loops until the sum of the
errors over all the intervals is below the required tolerance (Line 7). At
the top of the loop, the interval with the largest error is selected (Line 8).
If the error in this interval is below the numerical accuracy available for
the rule used or the interval is too small (*i.e.* the space between the first
two or last two nodes is zeros, Line 11), the interval is dropped and its

---

Algorithm 8 int_update (f, a, b, τ)

1: **for** $i = 0 \ldots n_{\text{init}}$ **do** $f_i \leftarrow f\left((a+b)/2 - x_i(a-b)/2\right)$ **end for**
   (*evaluate the integrand at the nodes x_i*)
2: $\mathbf{c} \leftarrow \mathbf{P}^{-1}\mathbf{f}$ (*compute the interpolation coefficients*)
3: $q_0 \leftarrow (b-a)c_0/\sqrt{2}$ (*approximate the integral*)
4: $\varepsilon_0 \leftarrow \infty$ (*start with a pessimistic error estimate*)
5: $H \leftarrow \{[a, b, \mathbf{c}, \mathbf{b}_{\text{def}}, q_0, \varepsilon_0, n_{\text{init}}, 0]\}$ (*init the heap with the first interval*)
6: $\varepsilon_{\text{xs}} \leftarrow 0$, $q_{\text{xs}} \leftarrow 0$ (*init the excess error and integral*)
7: **while** $\sum_{\varepsilon_i \in H} \varepsilon_i > \tau$ **do**
8:    $k \leftarrow \arg\max_k \varepsilon_k$ (*get the index of the interval with the largest error*)
9:    $H \leftarrow H \setminus \{[a_k, b_k, \mathbf{c}^{\text{old}}, \mathbf{b}^{\text{old}}, n_{\text{old}}, q_k, \varepsilon_k, n_{\text{div}}^{\text{old}}]\}$ (*remove the kth interval*)
10:   $h \leftarrow (b_k - a_k)/2$, $m \leftarrow (b_k + a_k)/2$ (*get the interval dimensions*)
11:   **for** half $\in \{\text{left}, \text{right}\}$ **do**
12:      $\mathbf{c}_{\text{old}}^{\text{half}} = \mathbf{T}^{\text{half}}\mathbf{c}^{\text{old}}$, $\mathbf{c}_{\text{half}} \leftarrow \mathbf{c}_{\text{old}}^{\text{half}}$ (*transform $\mathbf{c}$ from the previous level*)
13:      $\mathbf{b}_{\text{old}}^{\text{half}} = \mathbf{T}^{\text{half}}\mathbf{b}^{\text{old}}$, $\mathbf{b}_{\text{half}} \leftarrow \mathbf{b}_{\text{old}}^{\text{half}}$ (*transform $\mathbf{b}$ from the previous level*)
14:      $n_{\text{half}} \leftarrow n_{\text{old}}$ (*inherit the degree of the previous level*)
15:      **for** $i = 1 \ldots n_{\text{init}}/2$ **do**
16:        $f_i \leftarrow f(m + hx_i^{\text{half}})$ (*evaluate the integrand*)
17:        **if** $f_i \notin \{\text{NaN}, \pm\text{Inf}\}$ **then**
18:          $v_0 \leftarrow 1, v_1 \leftarrow x_i^{\text{half}}$ (*evaluate the polynomials at x_i^{half}*)
19:          **for** $j = 1 \ldots n_{\text{half}}$ **do** $v_j \leftarrow (x_i^{\text{half}} - \beta_j)v_{j-1} - \gamma_j v_{j-2}$ **end for**
20:          $\mathbf{c}_{\text{half}} \leftarrow \mathbf{c}_{\text{half}} + \dfrac{\mathbf{v}^\mathsf{T}\mathbf{c}_{\text{half}} - f_i}{\mathbf{v}^\mathsf{T}\mathbf{b}_{\text{half}}}\mathbf{b}_{\text{half}}$ (*update $\mathbf{c}$ as per Equation 3.33*)
21:          $\mathbf{b}_{\text{half}} \leftarrow (\mathbf{T} - \mathbf{I}x_i^{\text{half}})\mathbf{b}_{\text{half}}$ (*update $\mathbf{b}$ as per Equation 3.34*)
22:          $n_{\text{half}} \leftarrow n_{\text{half}} + 1$ (*increase the degree*)
23:        **end if**
24:      **end for**
25:      **if** $n_{\text{half}} > n_{\text{max}} \vee \frac{\|\mathbf{c}_{\text{half}} - \mathbf{c}_{\text{old}}\|}{\|\mathbf{c}_{\text{half}}\|} > \text{hint}$ **then**
26:        $\mathbf{c}_{\text{half}} \leftarrow \mathbf{P}^{-1}\mathbf{f}_{\text{half}}$, $\mathbf{b}_{\text{half}} \leftarrow \mathbf{b}_{\text{def}}$, $n_{\text{half}} \leftarrow n_{\text{init}}$ (*reset the interpolation*)
27:      **end if**
28:      $q_{\text{half}} \leftarrow hc_0^{\text{half}}/\sqrt{2}$ (*approximate the new integral*)
29:      $\varepsilon_{\text{half}} \leftarrow h\|\mathbf{c}_{\text{half}} - \mathbf{c}_{\text{old}}\|$ (*approximate the new error*)
30:      **if** $q_{\text{half}} \geq q_k^{(0)}$ **then** $n_{\text{div}}^{\text{half}} \leftarrow n_{\text{div}}^{\text{old}} + 1$ **else** $n_{\text{div}}^{\text{half}} \leftarrow n_{\text{div}}^{\text{old}}$ **end if**
31:      **if** $n_{\text{div}}^{\text{half}} > n_{\text{divmax}} \wedge 2n_{\text{div}}^{\text{half}} >$ rec. depth **then return** error **end if**
   (*abort on divergence*)
32:      **if** $\varepsilon_k < |q_k|\varepsilon_{\text{mach}}\text{cond}(\mathbf{P}) \vee$ interval too small **then**
33:        $H \leftarrow H \cup \{[a_{k+1}, b_{k+1}, \mathbf{c}_{\text{half}}, \mathbf{b}_{\text{half}}, q_{\text{half}}, \varepsilon_{\text{half}}, n_{\text{div}}^{\text{half}}]\}$
   (*push the new interval back on the heap*)
34:      **else**
35:        $\varepsilon_{\text{xs}} \leftarrow \varepsilon_{\text{xs}} + \varepsilon_k$, $q_{\text{xs}} \leftarrow q_{\text{xs}} + q_k$ (*collect the excess error and integral*)
36:      **end if**
37:    **end for**
38: **end while**
39: **return** $q_{\text{xs}} + \sum_{q_i \in H} q_i, \varepsilon_{\text{xs}} + \sum_{\varepsilon_i \in H} \varepsilon_i$ (*return the integral and the error*)

Algorithm 9 int_naive $\left(f, a, b, \tau\right)$

1: **for** $i = 0 \ldots n_{d_{\max}}$ **do** $f_i \leftarrow f\left((a+b)/2 - (a-b)x_i^{(m)}/2\right)$ **end for**

                                                                    (evaluate the integrand at the nodes $x_i^{(m)}$)

2: $\mathbf{c}^{(d_{\max}-1)} \leftarrow (\mathbf{P}^{(d_{\max}-1)})^{-1}\mathbf{f}[1:2:n_m+1]$, $\mathbf{c}^{(d_{\max})} \leftarrow (\mathbf{P}^{(d_{\max})})^{-1}\mathbf{f}$

                                                   (compute the interpolation coefficients)

3: $q_0 \leftarrow (b-a)c_0^{(d_{\max})}/\sqrt{2}$ *(approximate the integral)*

4: $\varepsilon_0 \leftarrow (b-a)\|\mathbf{c}^{(d_{\max})} - \mathbf{c}^{(d_{\max}-1)}\|$ *(approximate the error)*

5: $H \leftarrow \{[a, b, \mathbf{c}^{(d_{\max})}, q_0, \varepsilon_0, d_{\max}, 0]\}$ *(init the heap with the first interval)*

6: $\varepsilon_{xs} \leftarrow 0$, $q_{xs} \leftarrow 0$ *(init the excess error and integral)*

7: **while** $\sum_{\varepsilon_i \in H} \varepsilon_i > \tau$ **do**

8:     $k \leftarrow \arg\max_k \varepsilon_k$ *(get the index of the interval with the largest error)*

9:     $m \leftarrow (a+b)/2$, $h \leftarrow (b-a)/2$ *(get dimensions of the interval)*

10:    split $\leftarrow$ **false**   *(init split)*

11:    **if** $\varepsilon_k < |q_k|\varepsilon_{\mathrm{mach}}\mathrm{cond}(\mathbf{P}^{(d_k)}) \vee$ interval too small **then**

12:       $\varepsilon_{xs} \leftarrow \varepsilon_{xs} + \varepsilon_k$, $q_{xs} \leftarrow q_{xs} + q_k$ *(collect the excess error and integral)*

13:       $H \leftarrow H \setminus \{[a_k, b_k, \mathbf{c}^{old}, q_k, \varepsilon_k, d_k]\}$ *(remove the kth interval)*

14:    **else if** $d_k < d_{\max}$ **then**

15:       $d_k \leftarrow d_k + 1$ *(increase the degree in this interval)*

16:       **for** $i = 0 \ldots n_{d_k}$ **do** $f_i \leftarrow f(m + hx_i^{(d_k)}/2)$ **end for**

                                             (evaluate the integrand at the nodes $x_i^{(d_k)}$)

17:       $\mathbf{c}^{(d_k)} = (\mathbf{P}^{(d_k)})^{-1}\mathbf{f}$ *(compute the new interpolation coefficients)*

18:       $q_k \leftarrow (b_k - a_k)c_0^{(d_k)}/\sqrt{2}$ *(approximate the new integral)*

19:       $\varepsilon_k \leftarrow (b_k - a_k)\|\mathbf{c}^{(d_k)} - \mathbf{c}^{old}\|$ *(approximate the new error)*

20:       split $\leftarrow \dfrac{\|\mathbf{c}^{(d_k)} - \mathbf{c}^{old}\|}{\|\mathbf{c}^{(d_k)}\|} >$ hint   *(check change in the coefficients)*

21:    **else**

22:       split $\leftarrow$ **true**   *(split the interval if we are already at highest-degree rule)*

23:    **end if**

24:    **if** split **then**

25:       $H \leftarrow H \setminus \{[a_k, b_k, \mathbf{c}^{old}, q_k, \varepsilon_k, d_k, n_{div}^{old}]\}$ *(remove the kth interval)*

26:       **for** $i = 0 \ldots n_0$ **do**

27:       $f_i^{left} \leftarrow f\left((a+m)/2 + hx_i^{(0)}/2\right)$, $f_i^{right} \leftarrow f\left((m+b)/2 + hx_i^{(0)}/2\right)$

                               (evaluate the integrand at the nodes x_i^0 in the sub-intervals)

28:       **end for**

29:       **for** half $\in \{\mathrm{left}, \mathrm{right}\}$ **do**

30:       $\mathbf{c}^{half} = (\mathbf{P}^{(0)})^{-1}\mathbf{f}^{half}$ *(compute the new interpolation coefficients)*

31:       $q_{half} \leftarrow hc_0^{half}/\sqrt{2}$ *(approximate the new integral)*

32:       **if** $q_{half} \geq q_k^{(0)}$ **then** $n_{div}^{half} \leftarrow n_{div}^{old} + 1$ **else** $n_{div}^{half} \leftarrow n_{div}^{old}$ **end if**

33:       **if** $n_{div}^{half} > n_{divmax} \wedge 2n_{div}^{half} >$ rec. depth **then return** error **end if**

                                               (abort on divergence)

34:       $\varepsilon_{half} \leftarrow h\|\mathbf{c}^{half} - \mathbf{T}^{half}\mathbf{c}^{old}\|$ *(approximate the new error)*

35:       **end for**

36:       $H \leftarrow H \cup \{[a_k, m, \mathbf{c}^{left}, q_{left}, \varepsilon_{left}, 0, n_{div}^{left}], [m, b_k, \mathbf{c}^{right}, q_{right}, \varepsilon_{right}, 0, n_{div}^{right}]\}$

                                        (push the new intervals back on the heap)

37:    **end if**

38: **end while**

39: **return** $q_{xs} + \sum_{q_i \in H} q_i, \varepsilon_{xs} + \sum_{\varepsilon_i \in H} \varepsilon_i$ *(return the integral and the error)*

error and integral are accumulated in the excess variables ε_{xs} and q_{xs}
(Line 12). If the selected interval has not already used the highest-degree
rule (Line 14), the coefficients of the next-higher degree rule are computed and the integral and error estimate are updated (Lines 15 to 19).
The interval is bisected if either the highest-degree rule has already been
applied or if when increasing the degree of the rule the coefficients change
too much (Line 20). For the two new sub-intervals, the coefficients for the
lowest-degree rule are computed (Line 30) and used to approximate the
integral (Line 31). The number of times the integral increases over the
sub-interval is counted in the variables n_{div} (Line 32) and if they exceed
n_{divmax} and half of the recursion depth of that interval, the algorithm
aborts (Line 33) as per Section 5.1. Note that since the test in Equation 4.28 requires that both estimates be of the same degree, we will use,
for the estimate $q_k^{(0)}$ from the parent interval, the estimate which was
computed for the 0th rule. The error estimate for the new interval is
computed by transforming the interpolation coefficients from the parent
interval using Equation 3.3 and using its difference to the interpolation
in the new interval (Line 34). When the sum of the errors falls below
the required tolerance, the algorithm returns its approximations to the
integral and the integration error (Line 39).

6.4 An Adaptive Quadrature Algorithm using the Refined Error Estimate

In this algorithm, we will use the refined error estimate from Section 4.2,
Equation 4.27. The error estimate is computed from the transformed
coefficients from the previous level of recursion and the coefficients computed in the current interval. Since both interpolations are of the same
degree, ε_2 from Equation 4.27 can be computed.

We use the Clenshaw-Curtis quadrature rules since they are closed (*i.e.* the
end-points are included) and have good interpolation properties as described in Section 6.1.2.

The algorithm is shown in Algorithm 10. In Lines 1 to 5 an initial estimate is computed and used to initialize the heap H. The error estimate
is set to ∞ since it can not be estimated (Line 4). While the sum of
error estimates is below the required tolerance, the algorithm selects the

Algorithm 10 int_refined (f, a, b, τ)

1: **for** $i = 0 \ldots n$ **do** $f_i \leftarrow f((a+b)/2 + x_i(b-a)/2)$ **end for**
                                                  (evaluate the integrand at the nodes x_i)
2: $\mathbf{c} \leftarrow \mathbf{P}^{-1}\mathbf{f}$,
                                                  (compute the interpolation coefficients)
3: $q_0 \leftarrow (b-a)c_0^{(d_{max})}/\sqrt{2}$ *(approximate the integral)*
4: $\varepsilon_0 \leftarrow \infty$ *(start with a somewhat negative appreciation of the error)*
5: $H \leftarrow \{[a, b, \mathbf{c}, \mathbf{b}, 0, q_0, \varepsilon_0, 0]\}$ *(init the heap with the first interval)*
6: $\varepsilon_{xs} \leftarrow 0, q_{xs} \leftarrow 0$ *(init the excess error and integral)*
7: **while** $\sum_{\varepsilon_i \in H} \varepsilon_i > \tau$ **do**
8:     $k \leftarrow \arg\max_k \varepsilon_k$ *(get the index of the interval with the largest error)*
9:     $H \leftarrow H \setminus \{[a_k, b_k, \mathbf{c}^{old}, \mathbf{b}^{old}, f_{old}^{(n+1)}, q_k, \varepsilon_k, n_{div}^{old}]\}$ *(remove the kth interval)*
10:     **if** $\varepsilon_k < |q_k|\varepsilon_{mach}\mathrm{cond}(\mathbf{P}) \vee$ interval too small **then**
11:         $\varepsilon_{xs} \leftarrow \varepsilon_{xs} + \varepsilon_k, q_{xs} \leftarrow q_{xs} + q_k$ *(collect the excess error and integral)*
12:     **else**
13:         $m \leftarrow (a+b)/2, h \leftarrow (b-a)/2$
14:         **for** $i = 0 \ldots n$ **do**
15:             $f_i^{left} \leftarrow f((a+m)/2 + hx_i/2), f_i^{right} \leftarrow f((m+b)/2 + hx_i/2)$
                         (evaluate the integrand at the nodes x_i in the sub-intervals)
16:         **end for**
17:         **for** half $\in \{\mathrm{left}, \mathrm{right}\}$ **do**
18:             $\mathbf{c}^{half} = (\mathbf{P})^{-1}\mathbf{f}^{half}$ *(compute the new interpolation coefficients)*
19:             $q_{half} \leftarrow hc_0^{half}/\sqrt{2}$ *(approximate the new integral)*
20:             **if** $q_{half} \geq q_k$ **then** $n_{div}^{half} \leftarrow n_{div}^{old} + 1$ **else** $n_{div}^{half} \leftarrow n_{div}^{old}$ **end if**
21:             **if** $n_{div}^{half} > n_{divmax} \wedge 2n_{div}^{half} >$ rec. depth **then return** error **end if**
                                         (abort on divergence)
22:             $f_{half}^{(n+1)} \leftarrow \frac{\|\mathbf{c}^{half} - \mathbf{T}^{(half)}\mathbf{c}^{old}\|}{\|\mathbf{b}^{half} - \mathbf{T}^{(half)}\mathbf{b}^{old}\|}$ *(approximate the higher derivative)*
23:             **if** $\max\left\{\left|\mathbf{Pc}^{old} - \mathbf{f}^{half}\right| - \vartheta_1 f_{half}^{(n+1)}\left|\mathbf{Pb}^{old}\right|\right\} > 0$ **then**
24:                 $\varepsilon_{half} \leftarrow h\|\mathbf{c}^{half} - \mathbf{c}^{old}\|$ *(compute the un-scaled error)*
25:             **else**
26:                 $\varepsilon_{half} \leftarrow hf_{half}^{(n+1)}\|\mathbf{b}^{half}\|$
                            (compute the extrapolated error (Equation 4.31))
27:             **end if**
28:         **end for**
29:         $H \leftarrow H \cup \{[a_k, m, \mathbf{c}^{left}, \mathbf{b}^{left}, f_{left}^{(n+1)}, q_{left}, \varepsilon_{left}, n_{div}^{left}], \ldots$
               $[m, b_k, \mathbf{c}^{right}, \mathbf{b}^{right}, f_{right}^{(n+1)}, q_{right}, \varepsilon_{right}, n_{div}^{left}]\}$
                            (push the new intervals back on the heap)
30:     **end if**
31: **end while**
32: **return** $q_{xs} + \sum_{q_i \in H} q_i, \varepsilon_{xs} + \sum_{\varepsilon_i \in H} \varepsilon_i$ *(return the integral and the error)*

interval with the largest error estimate (Line 8) and removes it from the heap (Line 9). As with the previous algorithm, if the error estimate is smaller than the numerical precision of the integral or the interval is too small, the interval is dropped (Line 10) and its error and integral estimates stored in the excess variables ε_{xs} and q_{xs} (Line 11). The algorithm then computes the new coefficients for each sub-intervals (Line 18), as well as their integral approximation (Line 19). If the integral over the sub-interval is larger than over the previous interval, the variable n_{div} is increased (Line 20) and if it exceeds n_{divmax} and half of the recursion depth, the algorithm aborts with an error (Line 21). In Line 23 the algorithm tests whether the conditions laid out in Equation 4.28 for the approximation of the $n+1$st derivative hold. If they do not, the un-scaled error estimate is returned (Line 24), otherwise, the scaled estimate is returned (Line 26). Finally, both sub-intervals are returned to the heap (Line 29). Once the required tolerance is met, the algorithm returns its approximations to the integral and the integration error (Line 32).

6.5 Numerical Tests

In the following, we will test the three algorithms described in earlier against the following routines:

- quadl, MATLAB's adaptive quadrature routine [67], based on Gander and Gautschi's adaptlob (see Section 2.20). This algorithm uses a 4-point Gauss-Lobatto quadrature rule and its 7-point Kronrod extension.

- DQAGS, QUADPACK's [83] adaptive quadrature routine based on de Doncker's adaptive extrapolation algorithm (see Section 2.14) using a 10-point Gauss quadrature rule and its 21-point Kronrod extension as well as the ϵ-Algorithm to extrapolate the integral and error estimate. The routine is called through the GNU Octave [16] package's quad routine.

- da2glob from [24], which uses a doubly-adaptive strategy over Newton-Cotes rules of degree 5, 9, 17 and 27 using the null rule error estimator described in Section 2.19.

$f(x)$	quad1			DQAGS			da2glob		
	✓	✗	n_{eval}	✓	✗	n_{eval}	✓	✗	n_{eval}
Eqn (6.3)	406 (1)	594	95	926	74	454	977	23	95
Eqn (6.4)	894	106	117	959	41	398	1000	0	59
Eqn (6.5)	868	132	45	998	2	179	1000	0	29
Eqn (6.6)	359	641	89	782	218 (49)	482	882	118	136
Eqn (6.7)	303	697	308	969	31 (4)	1692	997	3	480
Eqn (6.8)	994	6	746	1000	0	446	1000	0	523

$f(x)$	Algorithm 10			Algorithm 9			Algorithm 8		
	✓	✗	n_{eval}	✓	✗	n_{eval}	✓	✗	n_{eval}
Eqn (6.3)	1000	0	358	1000	0	280	990	10	211
Eqn (6.4)	1000	0	308	1000 (1)	0	175	1000	0	188
Eqn (6.5)	1000	0	100	1000	0	112	1000	0	67
Eqn (6.6)	1000	0	500	1000	0	342	1000	0	283
Eqn (6.7)	1000	0	1431	997	3	976	989	11	820
Eqn (6.8)	1000	0	685	1000	0	873	1000	0	562

Table 6.1: Results of the Lyness-Kaganove tests for $\tau = 10^{-3}$.

$f(x)$	quad1 ✓	X	n_{eval}	DQAGS ✓	X	n_{eval}	da2glob ✓	X	n_{eval}
Eqn (6.3)	398	602	368	908	92	1093	993 (7)	7	298
Eqn (6.4)	886	114	236	915	85	788	1000	0	100
Eqn (6.5)	770	230	103	985	15	365	1000	0	58
Eqn (6.6)	992	8	481	925 (1)	75 (75)	700	1000	0	290
Eqn (6.7)	982	18	1297	998	2 (2)	2031	1000	0	870
Eqn (6.8)	998	2 (2)	2029	1000	0	577	1000 (1)	0	695

$f(x)$	Algorithm 10 ✓	X	n_{eval}	Algorithm 9 ✓	X	n_{eval}	Algorithm 8 ✓	X	n_{eval}
Eqn (6.3)	1000	0	993	1000	0	866	1000	0	684
Eqn (6.4)	1000	0	627	1000	0	315	1000	0	414
Eqn (6.5)	1000	0	256	1000	0	313	1000	0	205
Eqn (6.6)	1000	0	766	1000	0	614	1000	0	488
Eqn (6.7)	1000	0	2258	1000	0	1811	1000	0	1483
Eqn (6.8)	1000	0	1188	1000	0	1196	1000	0	906

Table 6.2: Results of the Lyness-Kaganove tests for $\tau = 10^{-6}$.

$f(x)$	quad1 ✓	✗	n_{eval}	DQAGS ✓	✗	n_{eval}	da2glob ✓	✗	n_{eval}
Eqn (6.3)	367	633 (31)	1051	748 (68)	252 (151)	1784	874 (222)	126 (125)	627
Eqn (6.4)	886	114	356	869	131 (4)	1167	1000	0	141
Eqn (6.5)	780	220	184	979	21	567	1000	0	93
Eqn (6.6)	999	1	1211	924	76 (76)	824	1000	0	474
Eqn (6.7)	998	2	3336	998	2 (2)	2449	1000	0	1431
Eqn (6.8)	959	41 (41)	5186	1000 (2)	0	737	998 (63)	2 (2)	799

$f(x)$	Algorithm 10 ✓	✗	n_{eval}	Algorithm 9 ✓	✗	n_{eval}	Algorithm 8 ✓	✗	n_{eval}
Eqn (6.3)	900 (154)	100 (100)	2005	910 (26)	90 (88)	1849	916 (155)	84 (84)	1487
Eqn (6.4)	1000	0	946	1000	0	460	1000	0	725
Eqn (6.5)	1000	0	415	1000	0	520	1000	0	407
Eqn (6.6)	1000	0	1294	1000	0	1077	1000	0	898
Eqn (6.7)	1000	0	3853	1000	0	3254	1000	0	2730
Eqn (6.8)	1000 (3)	0	2025	1000 (2)	0	1393	1000 (6)	0	1480

Table 6.3: Results of the Lyness-Kaganove tests for $\tau = 10^{-9}$.

$f(x)$	quad1 ✓	quad1 ✗	n_{eval}	DQAGS ✓	DQAGS ✗	n_{eval}	da2glob ✓	da2glob ✗	n_{eval}
Eqn (6.3)	261	739 (375)	2843	473 (110)	527 (493)	2632	494 (267)	506 (506)	1068
Eqn (6.4)	879 (1)	121	488.10	828 (4)	172 (23)	1533.59	1000 (2)	0	184
Eqn (6.5)	795	205	315	962	38	766	1000	0	133
Eqn (6.6)	928	72	3217	866	134 (76)	995	996 (765)	4 (4)	710
Eqn (6.7)	741	259 (259)	8697	998	2 (2)	2988	1000 (849)	0	2068
Eqn (6.8)	249	751 (751)	9601	962 (546)	38 (38)	1213	548 (548)	452 (452)	894

$f(x)$	Algorithm 10 ✓	Algorithm 10 ✗	n_{eval}	Algorithm 9 ✓	Algorithm 9 ✗	n_{eval}	Algorithm 8 ✓	Algorithm 8 ✗	n_{eval}
Eqn (6.3)	566 (165)	434 (434)	9016	568 (103)	432 (432)	8593	601 (216)	399 (399)	2716
Eqn (6.4)	1000 (4)	0	1264	1000	0	606	1000 (9)	0	1140
Eqn (6.5)	1000	0	574	1000	0	736	1000 (2)	0	650
Eqn (6.6)	1000 (596)	0	10547	1000 (426)	0	18000	1000	0	1608
Eqn (6.7)	1000 (382)	0	9267	1000 (326)	0	12835	1000 (12)	0	4968
Eqn (6.8)	992 (644)	8 (8)	9039	994 (474)	6 (6)	20098	990 (889)	10 (10)	14689

Table 6.4: Results of the Lyness-Kaganove tests for $\tau = 10^{-12}$.

$f(x)$	quadl	DQAGS	da2glob	Algorithm 10	Algorithm 9	Algorithm 8
Eqn (6.3)	8.22 (86.37)	138.57 (304.55)	42.93 (417.35)	32.30 (90.01)	32.40 (115.66)	40.76 (192.31)
Eqn (6.4)	11.83 (100.97)	130.20 (327.03)	35.58 (596.84)	28.34 (91.87)	25.95 (147.97)	42.93 (227.95)
Eqn (6.5)	4.99 (109.92)	76.16 (424.25)	11.44 (392.45)	8.21 (81.98)	8.15 (72.28)	14.03 (207.66)
Eqn (6.6)	8.26 (92.12)	180.08 (373.35)	57.05 (417.98)	54.31 (108.45)	49.36 (144.27)	57.78 (203.83)
Eqn (6.7)	47.77 (154.72)	1152.31 (680.94)	186.14 (387.28)	172.28 (120.39)	193.79 (198.48)	243.97 (297.34)
Eqn (6.8)	64.38 (86.28)	152.31 (341.18)	116.72 (223.02)	56.95 (83.04)	84.49 (96.70)	100.19 (178.10)

$f(x)$	quadl	DQAGS	da2glob	Algorithm 10	Algorithm 9	Algorithm 8
Eqn (6.3)	27.41 (74.33)	312.34 (285.65)	76.48 (256.27)	82.18 (82.69)	69.93 (80.68)	111.51 (162.85)
Eqn (6.4)	18.55 (78.54)	243.77 (309.09)	48.41 (483.53)	59.72 (95.13)	47.14 (149.30)	73.38 (176.84)
Eqn (6.5)	9.96 (96.28)	116.52 (318.67)	27.31 (470.00)	20.87 (81.46)	25.62 (81.67)	40.78 (198.89)
Eqn (6.6)	43.85 (91.02)	256.97 (366.67)	78.52 (270.28)	74.85 (97.72)	58.91 (95.82)	92.14 (188.79)
Eqn (6.7)	203.37 (156.73)	1377.19 (678.02)	245.73 (282.45)	277.88 (123.03)	278.27 (153.59)	452.17 (304.82)
Eqn (6.8)	181.70 (89.53)	207.11 (358.81)	147.93 (212.84)	98.24 (82.66)	95.76 (80.04)	172.44 (190.27)

$f(x)$	quadl	DQAGS	da2glob	Algorithm 10	Algorithm 9	Algorithm 8
Eqn (6.3)	76.71 (72.97)	514.15 (288.07)	128.97 (205.56)	151.80 (75.69)	133.07 (71.95)	261.73 (176.01)
Eqn (6.4)	27.23 (76.37)	355.91 (304.85)	58.81 (414.71)	85.18 (90.04)	56.71 (123.12)	124.43 (171.43)
Eqn (6.5)	14.57 (78.77)	170.05 (299.75)	41.12 (438.68)	36.73 (88.49)	46.99 (90.28)	70.63 (173.31)
Eqn (6.6)	118.38 (97.68)	293.45 (356.00)	103.70 (218.40)	113.45 (87.65)	80.29 (74.54)	168.15 (187.16)
Eqn (6.7)	509.40 (152.70)	1620.46 (661.45)	306.04 (213.76)	488.79 (126.84)	425.67 (130.79)	863.07 (316.13)
Eqn (6.8)	469.28 (90.48)	263.71 (357.60)	153.95 (192.47)	168.82 (83.33)	113.38 (81.39)	306.77 (207.14)

$f(x)$	quadl	DQAGS	da2glob	Algorithm 10	Algorithm 9	Algorithm 8
Eqn (6.3)	207.77 (73.07)	757.52 (287.80)	165.00 (154.44)	755.60 (83.81)	800.29 (93.13)	535.59 (197.15)
Eqn (6.4)	35.95 (73.65)	455.38 (296.94)	65.30 (354.06)	100.82 (79.75)	68.09 (112.29)	190.77 (167.27)
Eqn (6.5)	24.05 (76.16)	228.15 (297.47)	42.37 (317.11)	47.29 (82.34)	57.66 (78.33)	98.66 (151.58)
Eqn (6.6)	277.48 (86.25)	349.05 (350.78)	120.99 (170.25)	999.03 (94.72)	2168.54 (120.47)	308.72 (191.95)
Eqn (6.7)	1257.42 (144.57)	1856.19 (621.07)	342.83 (165.75)	1180.90 (127.43)	1917.43 (149.39)	1616.21 (325.29)
Eqn (6.8)	793.89 (82.68)	399.50 (329.29)	153.20 (171.20)	797.74 (88.25)	2282.50 (113.57)	4022.79 (273.85)

Table 6.5: Average time in milliseconds per run and, in brackets, microseconds per evaluation of the integrand for the Lyness-Kaganove tests for $\tau = 10^{-3}$, 10^{-6}, 10^{-9} and 10^{-12}.

The three new algorithms were implemented in the MATLAB programming language and are listed in Appendix B.

Over the years, several authors have specified sets of test functions to evaluate the performance and reliability of quadrature routines. In the following, we will use, with some minor modifications, the test "families" suggested by Lyness and Kaganove [64] and the "battery" of functions compiled by Gander and Gautschi [33], which are an extension of the set proposed by Kahaner [49].

The function families used for the Lyness-Kaganove test are

$$\int_0^1 |x - \lambda|^\alpha \, dx, \qquad \lambda \in [0,1], \ \alpha \in [-0.5, 0] \quad (6.3)$$

$$\int_0^1 (x > \lambda) e^{\alpha x} \, dx, \qquad \lambda \in [0,1], \ \alpha \in [0,1] \quad (6.4)$$

$$\int_0^1 \exp(-\alpha|x - \lambda|) \, dx, \qquad \lambda \in [0,1], \ \alpha \in [0,4] \quad (6.5)$$

$$\int_1^2 10^\alpha / ((x - \lambda)^2 + 10^\alpha) \, dx, \qquad \lambda \in [1,2], \alpha \in [-6,-3] \quad (6.6)$$

$$\int_1^2 \sum_{i=1}^4 10^\alpha / ((x - \lambda_i)^2 + 10^\alpha) \, dx, \qquad \lambda_i \in [1,2], \alpha \in [-5,-3] \quad (6.7)$$

$$\int_0^1 2\beta(x - \lambda) \cos(\beta(x - \lambda)^2) \, dx, \qquad \lambda \in [0,1], \ \alpha \in [1.8, 2], \quad (6.8)$$

$$\beta = 10^\alpha / \max\{\lambda^2, (1 - \lambda)^2\}$$

where the Boolean expressions are evaluated to 0 or 1. The integrals were evaluated to relative precisions of $\tau = 10^{-3}$, 10^{-6}, 10^{-9} and 10^{-12} for 1 000 realizations of the random parameters λ and α. The results of these tests are shown in Tables 6.1 to 6.4. For each function, the number of correct and incorrect integrations is given with, in brackets, the number of cases each in which a warning (either explicit or if an error estimate larger than the requested tolerance is returned) was issued.

The functions used for the "battery" test are

$$f_1 = \int_0^1 e^x \, \mathrm{d}x$$

$$f_2 = \int_0^1 H(x - 0.3) \, \mathrm{d}x$$

$$f_3 = \int_0^1 x^{1/2} \, \mathrm{d}x$$

$$f_4 = \int_{-1}^1 (\tfrac{23}{25} \cosh(x) - \cos(x)) \, \mathrm{d}x$$

$$f_5 = \int_{-1}^1 (x^4 + x^2 + 0.9)^{-1} \, \mathrm{d}x$$

$$f_6 = \int_0^1 x^{3/2} \, \mathrm{d}x$$

$$f_7 = \int_0^1 x^{-1/2} \, \mathrm{d}x$$

$$f_8 = \int_0^1 (1 + x^4)^{-1} \, \mathrm{d}x$$

$$f_9 = \int_0^1 2(2 + \sin(10\pi x))^{-1} \, \mathrm{d}x$$

$$f_{10} = \int_0^1 (1 + x)^{-1} \, \mathrm{d}x$$

$$f_{11} = \int_0^1 (1 + e^x)^{-1} \, \mathrm{d}x$$

$$f_{12} = \int_0^1 x(e^x - 1)^{-1} \, \mathrm{d}x$$

$$f_{13} = \int_{0.1}^1 \sin(100\pi x)/(\pi x) \, \mathrm{d}x$$

$$f_{14} = \int_0^{10} \sqrt{50} e^{-50\pi x^2} \, \mathrm{d}x$$

$$f_{15} = \int_0^{10} 25 e^{-25x} \, \mathrm{d}x$$

$$f_{16} = \int_0^{10} 50(\pi(2500x^2 + 1))^{-1} \, \mathrm{d}x$$

$$f_{17} = \int_{0.01}^1 50(\sin(50\pi x)/(50\pi x))^2 \, \mathrm{d}x$$

$$f_{18} = \int_0^\pi \cos(\cos(x) + 3\sin(x) + 2\cos(2x) + 3\cos(3x)) \, \mathrm{d}x$$

$$f_{19} = \int_0^1 \log(x) \, \mathrm{d}x$$

$$f_{20} = \int_{-1}^1 (1.005 + x^2)^{-1} \, \mathrm{d}x$$

$$f_{21} = \int_0^1 \sum_{i=1}^3 \left[\cosh(20^i(x - 2i/10)) \right]^{-1} \, \mathrm{d}x$$

$$f_{22} = \int_0^1 4\pi^2 x \sin(20\pi x) \cos(2\pi x) \, \mathrm{d}x$$

$$f_{23} = \int_0^1 (1 + (230x - 30)^2)^{-1} \, \mathrm{d}x$$

$$f_{24} = \int_0^3 \lfloor e^x \rfloor \, \mathrm{d}x$$

$$f_{25} = \int_0^5 (x + 1)(x < 1) + (3 - x)(1 \le x \le 3) + 2(x > 3) \, \mathrm{d}x$$

where the Boolean expressions in f_2 and f_{25} evaluate to 0 or 1. The functions are taken from [33] with the following modifications:

- No special treatment is given to the case $x = 0$ in f_{12}, allowing the integrand to return NaN.

- f_{13} and f_{17} are integrated from 0 to 1 as opposed to 0.1 to one and 0.01 to 1 respectively, allowing the integrand to return NaN for $x = 0$.

- No special treatment of $x < 10^{-15}$ in f_{19} allowing the integrand to return $-$Inf.

- f_{24} was suggested by J. Waldvogel as a simple test function with multiple discontinuities.

- f_{25} was introduced in [33], yet not used in the battery test therein.

The rationale for the modifications of f_{12}, f_{13}, f_{17} and f_{19} is that we can't, on one hand, assume that the user was smart or diligent enough to remove the non-numerical values in his or her integrand, and on the other hand assume that he or she would still resort to a general-purpose quadrature routine to integrate it. Any general purpose quadrature routine should be robust enough to deal with any function, provided by either careful or careless users.

The changes have little effect on quadl and DQAGS since, as mentioned in Section 5.1, the former shifts the integration boundaries by ε_{mach} if a non-numerical value is encountered on the edges of the integration domain and the latter uses Gauss and Gauss-Kronrod quadrature rules which do not contain the end-points and thus avoid the NaN returned at $x = 0$ for f_{12}, f_{13} and f_{17} and the Inf at $x = 0$ in f_{19}. da2glob, however, only deals with $\pm$Inf explicitly, replacing such values with zero. To be able to deal with f_{12}, f_{13} and f_{17}, the algorithm was modified to treat NaN in the same way as Inf.

The battery functions were integrated for the relative tolerances $\tau = 10^{-3}$, 10^{-6}, 10^{-9} and 10^{-12}. The results are summarized in Table 6.6, where the number of function evaluations for each combination of integrand, integrator and tolerance are given. If integration was unsuccessful, the number is stricken through. If the number of evaluations was the lowest for the given integrand and tolerance, it is shown in bold face.

$f(x)$	$\tau = 10^{-3}$						$\tau = 10^{-6}$					
	quad1	DQAGS	da2glob	Alg. 10	Alg. 9	Alg. 8	quad1	DQAGS	da2glob	Alg. 10	Alg. 9	Alg. 8
f_1	18	21	**9**	27	33	17	18	21	**9**	27	33	17
f_2	108	357	**61**	283	161	185	198	357	**101**	603	301	409
f_3	48	105	**25**	107	101	49	108	231	**65**	315	429	177
f_4	18	21	**9**	27	33	17	18	21	**9**	27	33	17
f_5	18	21	**17**	27	33	25	48	**21**	33	59	95	65
f_6	18	21	**9**	27	33	17	48	105	**41**	123	159	65
f_7	289	231	**121**	411	269	329	**231**	231	285	1035	709	673
f_8	18	21	**17**	27	33	17	**18**	21	25	43	33	33
f_9	198	315	**121**	251	261	201	468	399	**233**	411	587	417
f_{10}	18	21	**9**	27	33	17	18	21	**17**	27	33	17
f_{11}	18	21	**9**	27	33	17	18	21	**9**	27	33	17
f_{12}	19	21	49	59	47	**17**	19	21	105	59	55	**17**
f_{13}	~~589~~	**651**	929	891	1403	1129	1519	**1323**	1469	1611	2347	2153
f_{14}	78	231	**45**	187	151	105	138	231	**65**	203	183	137
f_{15}	78	147	**41**	187	135	97	168	189	**69**	219	159	113
f_{16}	18	21	**9**	27	33	17	18	21	**9**	27	33	17
f_{17}	**79**	483	325	603	903	649	~~940~~	**777**	1065	1179	1491	1457
f_{18}	108	105	**73**	123	145	105	228	147	**129**	187	209	193
f_{19}	109	231	**65**	155	255	177	229	231	**145**	475	717	513
f_{20}	18	21	**17**	27	33	17	48	**21**	33	59	33	65
f_{21}	~~138~~	~~273~~	**85**	~~235~~	~~203~~	137	~~348~~	~~357~~	~~185~~	~~347~~	~~391~~	~~273~~
f_{22}	228	**147**	241	235	371	257	**228**	315	305	379	627	449
f_{23}	108	273	**93**	299	191	161	258	399	**161**	411	281	281
f_{24}	~~138~~	1911	**453**	3227	4515	1905	~~1878~~	~~3211~~	~~857~~	9163	11433	**6793**
f_{25}	108	567	**81**	379	277	225	348	819	**149**	859	593	569

Table 6.6: Results of battery test for $\tau = 10^{-3}$ and 10^{-6}. For each test, the best result (least function evaluations) is in bold and unsuccessful runs are stricken through.

$f(x)$	$\tau = 10^{-9}$						$\tau = 10^{-12}$					
	quad1	DQAGS	da2g_ob	Alg. 10	Alg. 9	Alg. 8	quad1	DQAGS	da2glob	Alg. 10	Alg. 9	Alg. 8
f_1	**18**	21	**9**	27	33	17	18	21	**17**	27	33	33
f_2	318	357	**141**	923	441	681	408	357	**181**	1243	581	1097
f_3	258	231	**137**	523	799	393	~~648~~	231	241	827	1191	721
f_4	**18**	21	25	27	33	33	48	**21**	33	27	33	33
f_5	**48**	63	65	59	95	97	168	63	65	123	219	129
f_6	108	189	**73**	251	359	169	288	189	**137**	427	607	337
f_7	~~889~~	231	581	1867	1409	1281	~~2429~~	231	965	3291	2179	2209
f_8	48	**21**	33	59	95	65	138	63	**49**	123	95	89
f_9	1038	567	**401**	891	991	769	2808	735	**577**	1387	1425	1449
f_{10}	48	21	**17**	59	33	33	48	**21**	33	75	33	65
f_{11}	18	21	**9**	27	33	17	48	**21**	**17**	43	33	33
f_{12}	19	21	217	59	63	**17**	**19**	**21**	557	59	63	33
f_{13}	4879	**1323**	1913	2939	2459	2993	~~10039~~	**1323**	2233	5019	2521	6081
f_{14}	228	273	**105**	315	225	185	588	273	**153**	379	369	345
f_{15}	288	189	**101**	283	191	161	708	231	**145**	379	277	337
f_{16}	18	21	**9**	27	33	17	18	21	**9**	27	33	17
f_{17}	2839	**1323**	1725	2107	2419	2849	6469	**1323**	2077	3387	2451	3681
f_{18}	738	189	**185**	379	395	257	1758	**273**	**273**	731	581	593
f_{19}	~~499~~	231	285	875	1323	1033	1369	231	449	1563	1943	1833
f_{20}	**48**	63	65	91	95	73	168	63	65	187	219	129
f_{21}	**1158**	441	~~273~~	1179	~~653~~	~~521~~	2748	~~525~~	**649**	1771	~~1839~~	~~1025~~
f_{22}	2508	**315**	385	699	627	513	5568	**315**	513	1291	627	2249
f_{23}	588	441	**241**	699	569	497	1608	483	**401**	1083	957	865
f_{24}	~~3738~~	~~12285~~	~~101~~	15275	18503	**14537**	~~5538~~	~~16359~~	~~1745~~	21147	25191	20001
f_{25}	528	819	**201**	1339	933	1041	678	819	**269**	1803	1253	1617

Table 6.6: Results of battery test for $\tau = 10^{-9}$ and 10^{-12}. For each test, the best result (least function evaluations) is in bold and unsuccessful runs are stricken through.

Finally, the integrators were tested on the problem

$$\int_0^1 |x - \lambda|^\alpha \, dx, \quad \lambda \in [0, 1]$$

for 100 realizations of the random parameter λ and different values of α for a relative[3] tolerance $\tau = 10^{-6}$. Since for $alpha \leq -1$, the integral diverges and can not be computed numerically, we are interested in the warnings or errors returned by the different quadrature routines. The results are shown in Table 6.7. For each integrator we give the number of successes and failures as well as, in brackets, the number of times each possible error or warning was returned. The different errors, for each integrator, are:

- quadl: (Min/Max/Inf)

  - Min: Minimum step size reached; singularity possible.
  - Max: Maximum function count exceeded; singularity likely.
  - Inf: Infinite or Not-a-Number function value encountered.

- DQAGS: (ier$_1$/ier$_2$/ier$_3$/ier$_4$/ier$_5$)

  - ier$_1$: Maximum number of subdivisions allowed has been achieved.
  - ier$_2$: The occurrence of roundoff error was detected, preventing the requested tolerance from being achieved. The error may be under-estimated.
  - ier$_3$: Extremely bad integrand behavior somewhere in the interval.
  - ier$_4$: The algorithm won't converge due to roundoff error detected in the extrapolation table. It is presumed that the requested tolerance cannot be achieved, and that the returned result is the best which can be obtained.
  - ier$_5$: The integral is probably divergent or slowly convergent.

- da2glob: (noise/min/max/sing)

  - noise: The requested tolerance is below the noise level of the problem. Required tolerance may not be met.

---

[3]For $\alpha \leq -1$ an *absolute* tolerance of $\tau = 10^{-6}$ was used.

- min: Interval too small. Required tolerance may not be met.

- max: Maximum number of function evaluations. Required tolerance may not be met.

- sing: Singularity probably detected. Required tolerance may not be met.

• Algorithms 8, 9 and 10: (err/div)

- err: The final error estimate is larger than the required tolerance.

- div: The integral is divergent.

Thus, the results for DQAGS at $\alpha = -0.8$ should be read as the algorithm returning 146 correct and 854 false results (requested tolerance not satisfied) and having returned the error ier_3 (bad integrand behavior) 58 times and the error ier_5 (probably divergent integral) 4 times.

6.6 Results

As can be seen from the results in Tables 6.1 to 6.4, all three algorithms are clearly more reliable than quadl and DQAGS. MATLAB's quadl performs best for high precision requirements (small tolerances, best results for $\tau = 10^{-9}$), yet still fails often without warning. QUADPACK's DQAGS does better for low precision requirements (large tolerances, best results for $\tau = 10^{-3}$), yet also fails often, more often than not with a warning.

Espelid's da2glob does significantly better, with only a few failures at $\tau = 10^{-3}$ (without warning) and a large number of failures for Equation 6.8 at $\tau = 10^{-12}$, albeit all of them with prior warning. The former are due to the error estimate not detecting specific features of the integrand due to the interpolating polynomial looking smooth, when it is, in fact, singular (see Figure 6.8). The latter were due to the integral estimates being affected by noise (most often in the 17-point rule), which was, in all cases, detected and warned against by the algorithm.

The new algorithms fail only for (6.3) at higher precisions since the integral becomes numerically impossible to evaluate (there are not sufficient machine numbers near the singularity to properly resolve the integrand),

α	quad1	DQAGS	da2glob	Algorithm 10	Algorithm 9	Algorithm 8
α = −0.1	48 / 52 (0/0/0)	96 / 4 (0/0/0/0)	98 / 2 (0/0/0)	100 / 0 (0/0)	100 / 0 (0/0)	100 / 0 (0/0)
α = −0.2	44 / 56 (0/0/0)	93 / 7 (0/0/0/0)	100 / 0 (0/0/0)	100 / 0 (0/0)	100 / 0 (0/0)	100 / 0 (0/0)
α = −0.3	36 / 64 (0/0/0)	86 / 14 (0/0/0/0)	98 / 2 (0/0/0)	100 / 0 (0/0)	100 / 0 (0/0)	100 / 0 (0/0)
α = −0.4	31 / 69 (0/0/0)	83 / 17 (0/0/0/0)	100 / 0 (0/0/0)	100 / 0 (0/0)	100 / 0 (0/0)	100 / 0 (0/0)
α = −0.5	22 / 78 (0/0/0)	72 / 28 (0/0/0/0)	98 / 2 (0/0/0)	100 / 0 (0/0)	100 / 0 (0/0)	100 / 0 (0/0)
α = −0.6	10 / 90 (0/0/0)	46 / 54 (0/0/39/0)	100 / 0 (0/0/28)	100 / 0 (82/0)	100 / 0 (0/0)	100 / 0 (0/0)
α = −0.7	0 / 100 (0/0/44)	0 / 100 (0/77/0/19)	100 / 0 (0/0/94)	100 / 0 (99/0)	100 / 0 (99/0)	100 / 0 (80/0)
α = −0.8	0 / 100 (2/0/90)	0 / 100 (0/73/1/26)	0 / 100 (0/0/100)	100 / 0 (100/0)	100 / 0 (99/1)	100 / 0 (100/0)
α = −0.9	0 / 100 (1/0/98)	0 / 100 (0/76/1/23)	0 / 100 (0/0/100)	100 / 0 (94/6)	100 / 0 (88/12)	100 / 0 (100/0)
α = −1.0	0 / 100 (1/0/99)	0 / 100 (0/55/5/40)	0 / 100 (0/0/100)	100 / 0 (25/75)	100 / 0 (54/46)	100 / 0 (100/0)
α = −1.1	0 / 100 (1/0/99)	0 / 100 (0/36/5/59)	0 / 100 (0/0/100)	100 / 0 (0/100)	100 / 0 (7/93)	100 / 0 (56/44)
α = −1.2	0 / 100 (0/90/10)	0 / 100 (0/3/7/90)	0 / 100 (0/0/100)	100 / 0 (0/100)	100 / 0 (1/99)	100 / 0 (25/75)
α = −1.3	0 / 100 (0/100/0)	0 / 100 (0/1/6/93)	0 / 100 (0/0/100)	100 / 0 (0/100)	100 / 0 (1/99)	100 / 0 (18/82)
α = −1.4	0 / 100 (0/100/0)	0 / 100 (0/0/15/85)	0 / 100 (0/0/100)	100 / 0 (0/100)	100 / 0 (0/100)	100 / 0 (14/86)
α = −1.5	0 / 100 (0/100/0)	0 / 100 (0/0/26/74)	0 / 100 (0/0/100)	100 / 0 (0/100)	100 / 0 (0/100)	100 / 0 (6/94)
α = −1.6	0 / 100 (0/100/0)	0 / 100 (0/0/28/72)	0 / 100 (0/0/100)	100 / 0 (0/100)	100 / 0 (0/100)	100 / 0 (2/98)
α = −1.7	0 / 100 (0/100/0)	0 / 100 (0/0/32/68)	0 / 100 (0/0/100)	100 / 0 (0/100)	100 / 0 (0/100)	100 / 0 (2/98)
α = −1.8	0 / 100 (0/100/0)	0 / 100 (0/0/35/65)	0 / 100 (0/0/100)	100 / 0 (0/100)	100 / 0 (0/100)	100 / 0 (2/98)
α = −1.9	0 / 100 (0/100/0)	0 / 100 (0/0/38/62)	0 / 100 (0/0/100)	100 / 0 (0/100)	100 / 0 (0/100)	100 / 0 (2/98)
α = −2.0	0 / 100 (0/100/0)	0 / 100 (0/0/44/56)	0 / 100 (0/0/100)	100 / 0 (0/100)	100 / 0 (0/100)	100 / 0 (2/98)

Table 6.7: Results of computing $\int_0^1 |x - \lambda|^\alpha \, dx$ for 100 realizations of $\lambda \in [0, 1]$ for different α. The numbers are the number of correct and incorrect integrations as well as the number of times (in brackets) the different errors or warnings of each algorithm were returned.

for which a warning is issued. This problem is shared by the other integrators as well.

Algorithm 8 and Algorithm 9 also fail a few times when integrating Equation 6.7 at $\tau = 10^{-3}$. In all such cases, one of the peaks was simply missed completely by the lower-degree rules, giving the appearance of a flat curve.

Algorithm 8 also failed a few times to evaluate Equation 6.3 for $\tau = 10^{-3}$. These were cases where the assumption of the naive error estimate (see Equation 4.19) does not hold and although the resulting error estimate is large, it is just below the required tolerance.

All three algorithms also failed a few times on Equation 6.8 in cases where the resulting integral was several orders of magnitude smaller than the function values themselves, making the required tolerance practically un-attainable.

Whereas the new algorithms out-perform the others in terms of reliability, they do so at a cost of a higher number of function evaluations. On average, Algorithm 8 and Algorithm 9 require about *twice as many* function evaluations as da2glob, whereas Algorithm 10 requires roughly *six times* as many.

This trend is also visible in the results of the "battery" test (Table 6.6). The new algorithms all fail on f_{21} for all but the highest and second-highest tolerances respectively, since the third peak at $x = 0.6$ is missed completely. Espelid's also does quite well, failing on f_{21} at the same tolerances and for the same reasons as the new algorithms and on f_{24} for $\tau < 10^{-3}$, using, however, in almost all cases, less function evaluations than the new algorithms.

It is interesting to note that quadl, DAQGS and da2glob all failed to integrate f_{24} for tolerances $\tau < 10^{-3}$. A closer look at the intervals that caused each algorithm to fail (see Figure 6.9) reveals that in all cases, multiple discontinuities in the same interval caused the error estimate to be accidentally small, leading the algorithms to erroneously assume convergence. This is not a particularity of the interval chosen: if we integrate

$$\int_0^\lambda \lfloor e^x \rfloor \, dx, \quad \lambda \in [2.5, 3.5]$$

for $1\,000$ realizations of the random parameter λ for $\tau = 10^{-6}$ using these

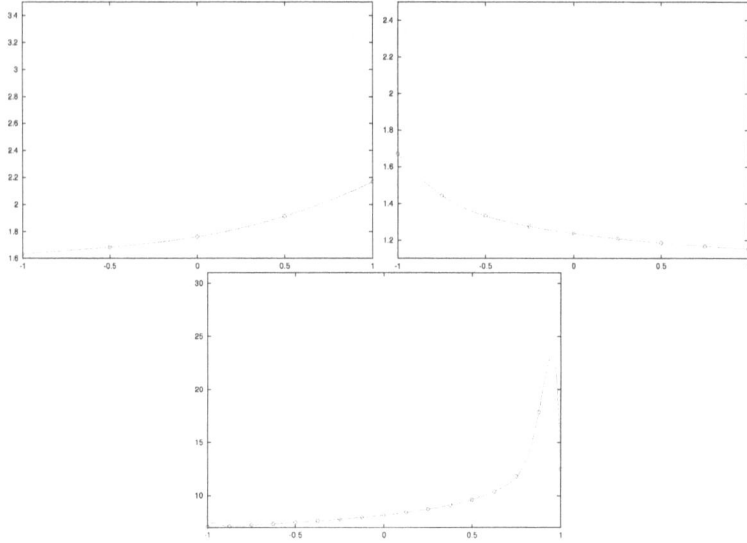

Figure 6.8: Some cases in which the 5, 9 and 17-point rules used in `da2glob` fail to detect a singularity (green). The assumed integrand (red) is sufficiently smooth such that its higher-degree coefficients, from which the error estimate is computed, are near zero.

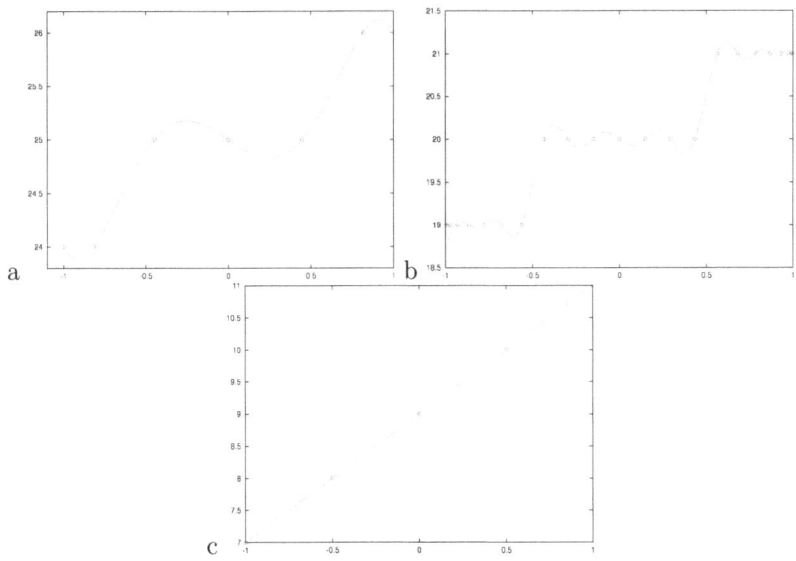

Figure 6.9: Intervals in which the error estimates of quadl (a), DQAGS (b) and da2glob (c) failed for f_{24}. The interpolatory polynomials used to compute the error estimates are shown in red.

three integrators, they fail on 894, 945 and 816 of the cases respectively. All three new algorithms succeed in all cases.

In the final test evaluating divergent integrals, (Table 6.7), quadl fails to distinguish between divergent and non-divergent integrals, reporting that a non-numerical value was encountered for $-1.1 \leq \alpha \leq -0.7$ and then aborting after the maximum 10 000 function evaluations[4] for $\alpha < 1.1$. For $\alpha < -1.0$, DQAGS reports the integral to be either subject to too much rounding error or divergent. The latter correct result was returned in more than half of the cases tested. In most cases where $\alpha < -0.6$, da2glob aborted, reporting that a singularity had probably been detected. This warning was usually accompanied by other warnings that the minimum interval size or the noise level had been reached (not shown).

---

[4]This termination criterion had been disabled for the previous tests.

For $\alpha = -0.8$, Algorithm 10 fails with a warning that the required tolerance was not met and as of $\alpha < -1.1$ both Algorithm 10 and Algorithm 9
abort, correctly, after deciding that the integral is divergent. Algorithm 8
is less successfull, reliably detecting the divergence only for $\alpha < -1.3$.

Chapter 7

Conclusions

In Chapter 3, we presented two new algorithms for the construction of polynomial interpolations. The first algorithm (Algorithm 4) offers no substantial improvement over that of Björck-Pereyra/Higham except that it can be easily downdated. The second algorithm, which does not allow for updates or downdates, is slightly more stable than the other algorithms tested and is more efficient when multiple right-hand sides need to be computed over the same set of nodes.

In Chapter 4, we presented two new error estimates. Over the set of problems treated therein, neither error estimate produced any false negative results, as opposed to the error estimators collected from Chapter 2.

In Chapter 6 we presented three new adaptive quadrature algorithms using the function representation discussed in Chapter 3, the error estimates described in Chapter 4 and the treatment of non-numerical function values and divergent integrals described in Chapter 5.

We conclude that the new algorithms presented in Chapter 6 are *more reliable* than MATLAB's `quadl`, QUADPACK's `DQAGS` and Espelid's `da2glob`. This increased reliability comes at a cost of a two to six times higher number of function evaluations required for complicated integrands.

The tradeoff between reliability and efficiency should, however, be of little concern in the context of automatic or general-purpose quadrature routines. Most modifications which increase efficiency usually rely on making certain assumptions on the integrand, *e.g.* smoothness, continuity, non-singularity, monotonically decaying coefficients, etc... If, however, the user knows enough about his or her integrand as to know that these assumptions are indeed valid and therefore that the algorithm will not fail, then he or she knows enough about the integrand as to *not* have to use a general-purpose quadrature routine and, if efficiency is crucial, should consider integrating it by a specialized routine or even analytically.

By making any assumptions for the user, we would be making two mistakes:

1. The increase in efficiency would *reward* users who, despite knowing enough about their integrand to trust the quadrature rule, have not made the effort to look for or implement a specialized routine,

2. The decrease in reliability *punishes* users who have turned to a general-purpose quadrature routine because they knowingly *can not* make any assumptions regarding their integrand.

It is for this reason that we should have no qualms whatsoever about sacrificing a bit of efficiency on certain integrands for much more reliability on tricky integrands of which we know, and can therefore assume, nothing.

Appendix A

Error Estimators in Matlab

A.1 Kuncir's 1962 Error Estimate

```
function [ int , err ] = est_kuncir1962 ( f , a , b , tol )

    % the nodes
    xi = a + (b-a)*[ 0.0 0.25 0.5 0.75 1 ]';

    % get the function values
    fx = f(xi);

    % compute the simpson's rules
    s1 = (b-a) * [ 1 0 4 0 1 ] * fx / 6;
    s2 = (b-a) * [ 1 4 2 4 1 ] * fx / 12;

    % compute the integral
    int = s2;

    % compute the error estimate
    err = abs( s1 - s2 );

end
```

A.2 McKeeman's 1962 Error Estimate

```
function [ int , err ] = est_mckeeman1962 ( f , a , b , tol )

    % the nodes
    xi = a + (b-a)*[ 0 1/6 1/3 1/2 2/3 5/6 1 ]';

    % get the function values
    fx = f(xi);
```

```
      % compute the simpson's rules
10    s1 = (b-a) * [ 1 0 0   4 0 0 1 ] * fx / 6;
      s2 = (b-a) * [ 1 4 2 4 2 4 1 ] * fx / 18;

      % compute the integral
      int = s2;
15
      % compute the error estimate
      err = abs( s1 - s2 );

   end
```

A.3   McKeeman's 1963 Error Estimate

```
function [ int , err ] = est_mckeeman1963_7 ( f , a , b , tol )

      % the nodes
      xi = a + (b-a)*linspace(0,1,50)';
5
      % get the function values
      fx = f(xi);

      % compute the simpson's rules
10    w = [ 751/17280, 3577/17280, 49/640, 2989/17280, ...
            2989/17280, 49/640, 3577/17280, 751/17280 ];
      s1 = (b-a) * w * fx(1:7:50);
      w = [ w(1:end-1) w(1)*2 w(2:end-1) w(1)*2 w(2:end-1) ...
            w(1)*2 w(2:end-1) w(1)*2 w(2:end-1) w(1)*2 ...
15          w(2:end-1) w(1)*2 w(2:end) ];
      s2 = (b-a) * w * fx / sum(w);

      % compute the integral
      int = s2;
20
      % compute the error estimate
      err = abs( s1 - s2 );

   end
```

A.4   Gallaher's 1967 Error Estimate

```
function [ int , err ] = est_gallaher1967 ( f , a , b , tol )

      % the nodes
      h = (1/6 + rand(1)/3) * (b - a);
5     xi = (b+a)/2 - [ h/2 0 -h/2 ]';

      % get the function values
      fx = f(xi);

10    % compute the integral
      int = (fx(1) + fx(3)) * (b-a-h) / 2 + fx(2)*h;

      % compute the error estimate
      err = 14.6 * abs(f(1)-2*f(2)+f(3)) * (b - a - h) / 2;
15 end
```

A.5   Lyness' 1969 Error Estimate

```
function [ int , err ] = est_lyness1969 ( f , a , b , tol )
```

```
      % the nodes
      xi = a + (b-a)*[ 0 1/8 1/4 3/8 1/2 5/8 3/4 7/8 1 ]';
5
      % get the function values
      fx = f(xi);

      % compute the 'old' simpson's rules
10    s1 = (b-a) * [ 1 0 0 0 4 0 0 0 1 ] * fx / 6;
      s2 = (b-a) * [ 1 0 4 0 2 0 4 0 1 ] * fx / 12;

      % compute the 'old' error estimate
      err_old = abs( s1 - s2 ) / 15;
15
      % get the int and err left and right...
      s1_left = (b-a) * [ 1 0 4 0 1 0 0 0 0 ] * fx / 12;
      s2_left = (b-a) * [ 1 4 2 4 1 0 0 0 0 ] * fx / 24;
20    int_left = (16*s2_left - s1_left) / 15;
      err_left = abs(s1_left - s2_left) / 15;
      s1_right = (b-a) * [ 0 0 0 0 1 0 4 0 1 ] * fx / 12;
      s2_right = (b-a) * [ 0 0 0 0 1 4 2 4 1 ] * fx / 24;
      int_right = (16*s2_right - s1_right) / 15;
      err_right = abs(s1_right - s2_right) / 15;
25
      % the integral
      int = int_left + int_right;

      % the noise-protected err
30    if err_old > tol % && (err_old < err_left || err_old < err_right)
          err = err_old;
      else
          err = err_left + err_right;
      end;
35
  end
```

A.6   O'Hara and Smith's 1969 Error Estimate

```
function [ int , err ] = est_ohara1969 ( f , a , b , tol )

      % evaluate the nodes
      xi = a + (b-a) * linspace(0,1,9)';
5     fx = f(xi);

      % compute nc9 and nc5_2
      nc9 = (b-a) * [ 989 5888 -928 10496 -4540 10496 -928 5888 989 ] * fx / 28350;
      nc5_2 = (b-a) * [ 7 32 12 32 14 32 12 32 7 ] * fx / 180;
10
      % compute err_1
      int = nc9;
      err = abs(nc9 - nc5_2);
      if err > tol, return; end;
15
      % evaluate cc7_2
      xi = a + (b-a) * ([ (cos( [6:-1:0]'/6*pi )-1)/2 ; (cos( [5:-1:0]'/6*pi )+1)/2 ] + 1)/2;
      fx = f(xi);
      cc7_2 = (b-a) * [ 0.028571428571428571425 0.25396825396825396823 0.45714285714285714282 ...
20        0.52063492063492063487 0.45714285714285714282 0.25396825396825396823 ...
          2*0.028571428571428571425 0.25396825396825396823 0.45714285714285714282 ...
          0.52063492063492063487 0.45714285714285714282 0.25396825396825396823 ...
          0.028571428571428571425 ] * fx / 4;

25    % test err_2
      int = cc7_2;
      err = abs(nc9 - cc7_2);
      if err > tol, return; end;

30    % evaluate err_3
      err = (b-a) * 32 / 945 * (abs([ 1/2 -1 1 -1 1 -1 1/2 ] * fx(1:7)) + ...
```

```
            abs([ 1/2 -1 1 -1 1 -1 1/2 ] * fx(7:13)));

end
```

A.7   De Boor's 1971 Error Estimate

```
    function [ int , err ] = est_deboor1971 ( f , a , b , tol )

        % allocate the T-table
        T = zeros(10);
5       R = zeros(10);

        % fill-in the first entry
        T(1,1) = (f(a) + f(b)) / 2;

10      % main loop...
        for l=2:10

            % compute T(l,1)
            T(l,1) = [1/2 ones(1,2^(l-1)-1) 1/2 ] * f(linspace(a,b,2^(l-1)+1)') / 2^(l-1);
15
            % line?
            if l==2 && T(1,1) == T(2,1)
                int = T(2,1);
                xi = [0.71420053 , 0.34662815 , 0.843751 , 0.12633046]';
20              fa = f(a); fb = f(b);
                fxi = f(a + (b-a)*xi);
                for i=1:4
                    if abs(fxi(i) - (fa + xi(i)*(fb - fa))) > eps * (abs(fa) + abs(fb))/2
                        err = 2*tol;
25                      return;
                    end;
                end;
                err = 0;
                return;
30          end;

            % fill in the rest of the row
            for i=2:l
                T(l,i) = T(l,i-1) + (T(l,i-1) - T(l-1,i-1)) / (4^(i-1) - 1);
35          end;

            % compute the R_i
            for i=1:l-2
                R(i) = (T(l-1,i) - T(l-2,i)) / (T(l,i) - T(l-1,i));
40          end;

            % if "sufficiently smooth"...
            if abs(R(1)-4) <= 0.15
                for i=2:l
45                  int = (b-a) * T(l,i);
                    err = (b-a) * abs(T(l,i) - T(l,i-1));
                    if err < tol, return; end;
                end

50          % jump discontinuity?
            elseif abs(R(1)-2) <= 0.01
                int = (b-a) * T(l,1);
                err = (b-a) * abs(T(l,1) - T(l-1,1));
                return;
55
            % other singularity?
            elseif l > 2 && abs((R(1)-R_old)/R(1)) <= 0.1
                if R(1) < 1.1
                    int = (b-a) * T(l,1);
60                  err = 2 * tol;
                    return;
                end;
                if R(1) > 4.5, continue; end;
                for i=2:l-1
65                  for j=i+1:l
```

```
                    T(j,i) = T(j,i-1) - (T(j-1,i-1) - T(j,i-1))^2 / ...
                             (T(j-1,i-1) - 2*T(j-1,i-1) + T(j,i-1));
                 end;
              end;
70            int = (b-a) * T(1,l-1);
              err = (b-a) * abs( T(1,l-1) - T(1,l-2) );
              return;

           % too many rows?
75         elseif l == 6
              int = T(1,1);
              err = 2*tol;
              return;

80         end;

           % remember the last ratio
           R_old = R(1);

85      end;

        % if we've landed here, then we have failed...
        int = T(1,1);
        err = 2*tol;
90   end
```

A.8   Oliver's 1972 Error Estimate

```
     function [ int , err ] = est_oliver1972 ( f , a , b , tol )

        % degree to begin with
        n = 4; j = 1; j_max = 2;
5
        % table of convergence rates
        Kn = [ 0.272 0.606 0.811 0.908 0.455
               0.144 0.257 0.376 0.511 0.550
               0.243 0.366 0.449 0.522 0.667
10             0.283 0.468 0.565 0.624 0.780
               0.290 0.494 0.634 0.714 0.855
               0.292 0.499 0.644 0.745 0.0 ];

        % main loop
15      while (1)

           % construct the nodes and get the function values
           xi = a + (b-a) * (cos([n:-1:0]/n*pi)' + 1) / 2;
           fx = f(xi);
20
           % compute the coefficients (brute-force vandermonde)
           xi = cos([n:-1:0]/n*pi)';
           V = [ ones(size(xi)) , xi ];
           for i = 2:n
25            V = [ V , 2 * xi .* V(:,end) - V(:,end-1) ];
           end;
           c = V \ fx;

           % approximate the integral
30         int = 2*c(1);
           for i=3:2:n+1
              int = int - 2 * c(i) / (i - 2) / i;
           end;
           int = int / 2 * (b - a);
35
           % get the convergence rate of the coeffs
           if j > 1
              if c(n-1) == 0, c(n-1) = eps * int; end;
              if c(n-3) == 0, c(n-3) = eps * int; end;
40            if c(n-5) == 0, c(n-5) = eps * int; end;
              K = max( abs( [ c(n+1)/c(n-1) , c(n-1)/c(n-3) , c(n-3)/c(n-5) ] ) );
              An2 = K^3 * abs(c(n-3));
```

```
       else
           if c(n-1) == 0, c(n-1) = eps * int; end;
45         if c(n-3) == 0, c(n-3) = eps * int; end;
           K = max( abs( [ c(n+1)/c(n-1) , c(n-1)/c(n-3) ] ) );
           An2 = K^2 * abs(c(n-1));
       end

50     % find sigma such that K < Kn(j,sigma)
       sigma = 1;
       while sigma < 5 && K > Kn(j,sigma), sigma = sigma + 1; end;

       % do we have a sigma and thus an error estimate?
55     if sigma < 5
           sigma = 2^sigma;
           err = (b-a) * sigma * 16*n / (n^2 - 1) / (n^2 - 9) * An2;
           if j > 1, err = min( err , abs(int - int_old) ); end;
           if err < tol || j == j_max, return; end;
60     elseif j > 1
           err = abs(int - int_old);
           if err < tol || j == j_max, return; end;
       end;

65     % otherwise split
       int_old = int;
       n = 2 * n; j = j + 1;

       end % while
70 end
```

A.9   Rowland and Varol's 1972 Error Estimate

```
function [ int , err ] = est_rowland1972 ( f , a , b , tol )

   % the nodes
   xi = a + (b-a)*linspace(0,1,9)';
5
   % get the function values
   fx = f(xi);

   % compute the simpson's rules
10 s1 = (b-a) * [ 1 0 0 0 4 0 0 0 1 ] * fx / 6;
   s2 = (b-a) * [ 1 0 4 0 2 0 4 0 1 ] * fx / 12;
   s4 = (b-a) * [ 1 4 2 4 2 4 2 4 1 ] * fx / 24;

   % compute the integral
15 int = s4;

   % compute the error estimate
   if s1 ~= s2
       err = (s2 - s4)^2 / abs( s1 - s2 );
20 else
       err = abs(s2 - s4);
   end;

end
```

A.10   Garribba *et al.*'s 1978 Error Estimate

```
function [ int , err ] = est_garribba1978 ( f , a , b , tol )
```

```
   % the nodes and weights
5  xi =[-0.9324695142031520278l, -0.66120938646626451366, ...
       -0.23861918608319690863, 0.23861918608319690863, 0.66120938646626451366, ...
       0.9324695142031520278l]';
   w = [0.17132449237917034513, 0.36076157304813860846, 0.46791393457269104698, ...
       0.46791393457269104752, 0.36076157304813860751, 0.17132449237917034507];
10
   % get the function values
   fx = f(a + (b-a) * (xi + 1)/2);
   fl = f(a + (b-a) * (xi + 1)/4);
   fr = f(a + (b-a) * (xi + 3)/4);
15
   % compute both quadratures
   q1 = (b-a) * w * fx / sum(w);
   q2 = (b-a) * w * (fl + fr) / sum(w) / 2;

20 % compute the integral
   int = q2;

   % compute the error estimate
   err = abs( q1 - q2 ) / (2^12 - 1);
25
   end
```

A.11   Piessens' 1979 Error Estimate

```
   function [ int , err ] = est_piessens1979 ( f , a , b , tol )

   % the nodes
   xi = [ 0.9956571630258081e+00,    0.9739065285171717e+00, ...
5         0.9301574913557082e+00,    0.8650633666889845e+00, ...
          0.7808177265864169e+00,    0.6794095682990244e+00, ...
          0.5627571346686047e+00,    0.4333953941292472e+00, ...
          0.2943928627014602e+00,    0.1488743389816312e+00, ...
          0.0000000000000000e+00 ]';
10 xi = [ xi ; -xi(end-1:-1:1) ];

   % get the function values
   fx = f(a + (b-a) * (xi + 1)/2);

15 % the weights
   w = [ 0.6667134430868814e-01,    0.1494513491505806e+00, ...
         0.2190863625159820e+00,    0.2692667193099964e+00, ...
         0.2955242247147529e+00 ];
   w = [ w , w(end:-1:1) ] / 2;
20 wk = [ 0.1169463886737187e-01,    0.3255816230796473e-01, ...
          0.5475589657435200e-01,    0.7503967481091995e-01, ...
          0.9312545458369761e-01,    0.1093871588022976e+00, ...
          0.1234919762620659e+00,    0.1347092173114733e+00, ...
          0.1427759385770601e+00,    0.1477391049013385e+00, ...
25        0.1494455540029169e+00 ];
   wk = [ wk , wk(end-1:-1:1) ] / 2;

   % compute the quadratures
   g = (b-a) * w * fx(2:2:end-1);
30 gk = (b-a) * wk * fx;

   % compute the integral
   int = gk;

35 % compute the error estimate
   err = abs( g - gk );

   end
```

A.12   Patterson's 1979 Error Estimate

```
     function [ int , err ] = est_patterson1979 ( f , a , b , tol )
     %
     %   Shameles Matlab translation of Krogh and Van Snyder since
     %   Patterson's original algorithm is numerically unstable
5
        % *****      Data Statements      ***************************************

        fl = [ 0 , 2, 3, 5, 9,12,14, 1];
        fh = [ 0 , 2, 4, 8,16,17,17, 0];
10      kl = [ 1, 1, 1, 1, 1, 3, 5, 9, 5, 9,12];
        kh = [ 1, 2, 4, 8,16, 3, 6,17, 5, 9,17];
        kx = [ 0, 1, 2, 3, 4, 5,      8,     11];

        % init P for efficiency
15      P = zeros(305,1);

        % In the comments below, F(K,I) refers to the function value
        % computed for the I'th node of the K'th formula.  The abscissae and
        % weights are stored in order according to the distance from the
20      % boundary of the region, not from the center.  Since we store
        % 1 - |abscissa|, the first "node" coefficient for each formula is
        % the smallest.

        % Corrections, nodes and weights for the 3-point formula.
25
        % Correction for F(1,1).
        P(  1) = -.111111111111111111111e+00;
        % Node and weight for F(2,1).
        P(  2) = +.22540333075851662296e+00; P(  3) = +.55555555555555555556e+00;
30
        % Corrections, nodes and weights for the 7-point formula.

        % Corrections for F(1,1) and F(2,1).
        P(  4) = +.64720942140296979le-02; P(  5) = -.928968790944433705e-02;
35      % Nodes and weights for F(3,1-2).
        P(  6) = +.39508731291979716579e-01; P(  7) = +.10465622602646726519e+00;
        P(  8) = +.56575625065319744200e+00; P(  9) = +.40139741477596222291e+00;

        % Corrections, nodes and weights for the 15-point formula.
40
        % Corrections for F(1,1), F(2,1), F(3,1-2).
        P( 10) = +.52230468969616226e-04; P( 11) = +.17121030961750000e-03;
        P( 12) = -.724830016153892898e-03; P( 13) = -.7017801099209042e-04;
        % Nodes and weights for F(4,1-4).
45      P( 14) = +.61680367872449777899e-02; P( 15) = +.17001719629940260339e-01;
        P( 16) = +.11154076712774300110e+00; P( 17) = +.92927195315124537686e-01;
        P( 18) = +.37889705326277359705e+00; P( 19) = +.17151190913639138079e+00;
        P( 20) = +.77661331357103311837e+00; P( 21) = +.21915685840158749640e+00;

50      % Corrections, nodes and weights for the 31-point formula.

        % Corrections for F(1,1), F(2,1), F(3,1-2), F(4,1-4).
        P( 22) = +.682166534792e-08; P( 23) = +.12667409859336e-06;
        P( 24) = +.59565976367837165e-05; P( 25) = +.1392330106826e-07;
55      P( 26) = -.6629407564902392e-04; P( 27) = -.704395804282302e-06;
        P( 28) = -.34518205339241e-07; P( 29) = -.814486910996e-08;
        % Nodes and weights for F(5,1-8).
        P( 30) = +.90187503233240234038e-03; P( 31) = +.25447807915618744154e-02;
        P( 32) = +.18468850446259893130e-01; P( 33) = +.16446049854387810934e-01;
60      P( 34) = +.70345142570259943330e-01; P( 35) = +.35957103307129322097e-01;
        P( 36) = +.16327406183113126449e+00; P( 37) = +.56979509494123357412e-01;
        P( 38) = +.29750379350847292139e+00; P( 39) = +.76879620499003531043e-01;
        P( 40) = +.46868025635562437602e+00; P( 41) = +.93627109981264473617e-01;
        P( 42) = +.66884640674202316691e+00; P( 43) = +.10566989358023480974e+00;
65      P( 44) = +.88751105686681337425e+00; P( 45) = +.11195687302095345688e+00;

        % Corrections, nodes and weights for the 63-point formula.

        % Corrections for F(1,1), F(2,1), F(3,1-2), F(4,1-4), F(5,1-8).
70      P( 46) = +.371583e-15; P( 47) = +.21237877e-12;
        P( 48) = +.10522629388435e-08; P( 49) = +.1748029e-14;
        P( 50) = +.3475718983017160e-06; P( 51) = +.90312761725e-11;
        P( 52) = +.12558916e-13; P( 53) = +.54591e-15;
        P( 54) = -.72338395508691963e-05; P( 55) = -.169699579757977e-07;
75      P( 56) = -.854363907155e-10; P( 57) = -.12281300930e-11;
```

```
     P( 58) = -.462334825e-13; P( 59) = -.42244055e-14;
     P( 60) = -.88501e-15; P( 61) = -.40904e-15;
     % Nodes and weights for F(6,1-16).
     P( 62) = +.12711187964238806027e-03; P( 63) = +.36322148184553065969e-03;
80   P( 64) = +.2793740627780409196e-02; P( 65) = +.25790497946856882724e-02;
     P( 66) = +.11315242452570520059e-01; P( 67) = +.61155068221172463397e-02;
     P( 68) = +.27817125251418203419e-01; P( 69) = +.10498246909621321898e-01;
     P( 70) = +.53657141626597094849e-01; P( 71) = +.15406750466559497802e-01;
     P( 72) = +.89628843042995707499e-01; P( 73) = +.20594233915912711149e-01;
85   P( 74) = +.13609206180630952284e+00; P( 75) = +.25869679327214746911e-01;
     P( 76) = +.19305946804978238813e+00; P( 77) = +.31073551111687964880e-01;
     P( 78) = +.26024395564730524132e+00; P( 79) = +.36064432780782572640e-01;
     P( 80) = +.33709033997521940454e+00; P( 81) = +.40715510116944318934e-01;
     P( 82) = +.42280428994795418516e+00; P( 83) = +.449145316353632197414e-01;
90   P( 84) = +.51638197305415897244e+00; P( 85) = +.48564330406673198716e-01;
     P( 86) = +.61664067580126965307e+00; P( 87) = +.51583253952048458777e-01;
     P( 88) = +.72225017797817568492e+00; P( 89) = +.53905499335266063927e-01;
     P( 90) = +.83176474844779253501e+00; P( 91) = +.55481404356559363988e-01;
     P( 92) = +.94365568695340721002e+00; P( 93) = +.56277699831254301273e-01;
95
     % Corrections, nodes and weights for the 127-point formula.

     % Corrections for F(3,1), F(4,1-2), F(5,1-3), F(6,1-6).
     P( 94) = +.1041098e-15; P( 95) = +.249472054598e-10;
100  P( 96) = +.55e-20; P( 97) = +.290412475995385e-07;
     P( 98) = +.367282126e-13; P( 99) = +.5568e-18;
     P(100) = -.871176477376972025e-06; P(101) = -.8147324267441e-09;
     P(102) = -.8830920337e-12; P(103) = -.18018239e-14;
     P(104) = -.70528e-17; P(105) = -.506e-19;
105  % Nodes and weights for F(7,1-32).
     P(106) = +.17569645108401419961e-04; P(107) = +.50536095207862517625e-04;
     P(108) = +.40120032808931675009e-03; P(109) = +.37774664632698460627e-03;
     P(110) = +.16833646815926074696e-02; P(111) = +.93836984854238150079e-03;
     P(112) = +.42758953015928114900e-02; P(113) = +.16811428654214699063e-02;
110  P(114) = +.85042788218938676006e-02; P(115) = +.25687649437940203731e-02;
     P(116) = +.14628500401479628890e-01; P(117) = +.35728927835172996449e-02;
     P(118) = +.22858485360294285840e-01; P(119) = +.46710503721143217474e-02;
     P(120) = +.33362148441583432910e-01; P(121) = +.58434498758356395076e-02;
     P(122) = +.46269993574238863589e-01; P(123) = +.70724899954335554680e-02;
115  P(124) = +.61679602220407116350e-01; P(125) = +.83428387539681577056e-02;
     P(126) = +.79659974529987579270e-01; P(127) = +.96411777297025366958e-02;
     P(128) = +.10025510022305996335e+00; P(129) = +.10955733387837901648e-01;
     P(130) = +.12348658551529473026e+00; P(131) = +.12275830560082770087e-01;
120  P(132) = +.14935550523164972024e+00; P(133) = +.13591571009765546790e-01;
     P(134) = +.17784374563501959262e+00; P(135) = +.14893641664815182035e-01;
     P(136) = +.20891506620015163857e+00; P(137) = +.16173218729577719942e-01;
     P(138) = +.24251603361948636206e+00; P(139) = +.17421930159464173747e-01;
     P(140) = +.27857691462990108452e+00; P(141) = +.18631848265138790186e-01;
125  P(142) = +.31701256890892077191e+00; P(143) = +.19795495048097499488e-01;
     P(144) = +.35772335749024048622e+00; P(145) = +.20905851445812023352e-01;
     P(146) = +.40059606975775100702e+00; P(147) = +.21956366305317824939e-01;
     P(148) = +.44550486736806745112e+00; P(149) = +.22940964229387748761e-01;
     P(150) = +.49231222424662833978e+00; P(151) = +.23854052106038540080e-01;
130  P(152) = +.54086998801010676712e+00; P(153) = +.24690524744487676909e-01;
     P(154) = +.59102017877011132759e+00; P(155) = +.25445769965464676581e-01;
     P(156) = +.64259616216846784762e+00; P(157) = +.2611567337670097680e-01;
     P(158) = +.69542355844328595666e+00; P(159) = +.266966229274050359906e-01;
     P(160) = +.74932126969651682339e+00; P(161) = +.27185533229624791819e-01;
135  P(162) = +.80410249728889984607e+00; P(163) = +.27579749566481873035e-01;
     P(164) = +.85957576684743982540e+00; P(165) = +.27877251476613701609e-01;
     P(166) = +.91554595991628911629e+00; P(167) = +.28076455793817246607e-01;
     P(168) = +.97181535105025430566e+00; P(169) = +.28176319033016602131e-01;

140  % Corrections, nodes and weights for the 255-point formula.

     % Corrections for F(4,1), F(5,1), F(6,1-2), F(7,1-4).
     P(170) = +.3326e-18; P(171) = +.114094770478e-11;
     P(172) = +.2952436056970351e-08; P(173) = +.51608328e-15;
145  P(174) = -.110177219650597323e-06; P(175) = -.58656987416475e-10;
     P(176) = -.23340340645e-13; P(177) = -.1248950e-16;
     % Nodes and weights for F(8,1-64).
     P(178) = +.240362020515353807630e-05; P(179) = +.69379364324108267170e-05;
     P(180) = +.56003792945624240417e-04; P(181) = +.5327529369780613125e-04;
     P(182) = +.23950907556795267013e-03; P(183) = +.13575491094922871973e-03;
150  P(184) = +.61966197497641806982e-03; P(185) = +.24921240048299729402e-03;
```

```
     P(186) = +.12543855319048853002e-02;  P(187) = +.38974528447328229322e-03;
     P(188) = +.21946455040427254399e-02;  P(189) = +.55429531493037471492e-03;
     P(190) = +.34858540851097261500e-02;  P(191) = +.74028280424450333046e-03;
     P(192) = +.51684971993789994803e-02;  P(193) = +.94536151685852538246e-03;
155  P(194) = +.72786557172113846706e-02;  P(195) = +.11674841174299594077e-02;
     P(196) = +.98486295992298408193e-02;  P(197) = +.14049079956551446427e-02;
     P(198) = +.12907472045965932809e-01;  P(199) = +.16561127281544526052e-02;
     P(200) = +.16481342421367271240e-01;  P(201) = +.19197129710138724125e-02;
     P(202) = +.20593718329137316189e-01;  P(203) = +.21944069253638388388e-02;
160  P(204) = +.25265540247597332240e-01;  P(205) = +.24789582266575679307e-02;
     P(206) = +.30515340497540768229e-01;  P(207) = +.27721957645934509940e-02;
     P(208) = +.36359378430187867480e-01;  P(209) = +.30730184347025783234e-02;
     P(210) = +.42811783890139037259e-01;  P(211) = +.33803979910869203823e-02;
     P(212) = +.49884702478705123440e-01;  P(213) = +.36933779170256508183e-02;
165  P(214) = +.57588434808916940190e-01;  P(215) = +.40110687240750233989e-02;
     P(216) = +.65931563842274211999e-01;  P(217) = +.43326409680929828545e-02;
     P(218) = +.74921067092924347640e-01;  P(219) = +.46573172997568547773e-02;
     P(220) = +.84562412844234959360e-01;  P(221) = +.49843645647655386012e-02;
     P(222) = +.94859641186738404810e-01;  P(223) = +.53130866051870565663e-02;
170  P(224) = +.10581543166444097714e+00;  P(225) = +.56428181013844441585e-02;
     P(226) = +.11743115975265809315e+00;  P(227) = +.59729195655081658049e-02;
     P(228) = +.12970694445186609414e+00;  P(229) = +.63027734490857587172e-02;
     P(230) = +.14264168911376784347e+00;  P(231) = +.66317812429018878941e-02;
     P(232) = +.15623311732729139895e+00;  P(233) = +.69593614093904229394e-02;
175  P(234) = +.17047780536259859981e+00;  P(235) = +.72849479805538070639e-02;
     P(236) = +.18537121234486258656e+00;  P(237) = +.76079896657190565832e-02;
     P(238) = +.20090770903915859819e+00;  P(239) = +.79279493342948491103e-02;
     P(240) = +.21708060588171698360e+00;  P(241) = +.82443037630328680306e-02;
     P(242) = +.23388218069623990928e+00;  P(243) = +.85565435613076896192e-02;
180  P(244) = +.25130370638306339718e+00;  P(245) = +.88641732094824942641e-02;
     P(246) = +.26933547875781873867e+00;  P(247) = +.91667111635607884067e-02;
     P(248) = +.28796684463774796540e+00;  P(249) = +.94636899938300652943e-02;
     P(250) = +.30718623022088529711e+00;  P(251) = +.97546565363174114611e-02;
     P(252) = +.32698116769768152079e+00;  P(253) = +.10039172044056840798e-01;
185  P(254) = +.34733833458998250389e+00;  P(255) = +.10316812330947621682e-01;
     P(256) = +.36824356228880576959e+00;  P(257) = +.10587167904885197931e-01;
     P(258) = +.38968188628481359983e+00;  P(259) = +.10849844089337314099e-01;
     P(260) = +.41163756555233745857e+00;  P(261) = +.11104461134006926537e-01;
     P(262) = +.43409411457634557737e+00;  P(263) = +.11350654315980596602e-01;
190  P(264) = +.45703433350168850951e+00;  P(265) = +.11588074033043952568e-01;
     P(266) = +.48044033846254297801e+00;  P(267) = +.11816385890830235763e-01;
     P(268) = +.50429359208123853983e+00;  P(269) = +.12035270785279562630e-01;
     P(270) = +.52857493412834112307e+00;  P(271) = +.12244424981611985899e-01;
     P(272) = +.55326461233797152625e+00;  P(273) = +.12443560190714035263e-01;
195  P(274) = +.57834231337383669993e+00;  P(275) = +.12632403643542078765e-01;
     P(276) = +.60378719394238406082e+00;  P(277) = +.12810698163877361967e-01;
     P(278) = +.62957791204992176986e+00;  P(279) = +.12978202239537399286e-01;
     P(280) = +.65569265584005619772e+00;  P(281) = +.13134690091960152836e-01;
     P(282) = +.68210918779315233168e+00;  P(283) = +.13279951743930530650e-01;
200  P(284) = +.70880485148175331803e+00;  P(285) = +.13413793085110098513e-01;
     P(286) = +.73575662758907323806e+00;  P(287) = +.13536035934956213614e-01;
     P(288) = +.76294115441017027278e+00;  P(289) = +.13646581810257129142e-01;
     P(290) = +.79033476175681880523e+00;  P(291) = +.13745093443001896632e-01;
205  P(292) = +.81791350324074780175e+00;  P(293) = +.13831631909506428676e-01;
     P(294) = +.84565318851862189130e+00;  P(295) = +.13906019601325461264e-01;
     P(296) = +.87352941562769803314e+00;  P(297) = +.13968158806516938516e-01;
     P(298) = +.90151760340188079791e+00;  P(299) = +.14017968039456608810e-01;
     P(300) = +.92959302395714482093e+00;  P(301) = +.14055382072649964277e-01;
210  P(302) = +.95773083523463639678e+00;  P(303) = +.14080351962553661325e-01;
     P(304) = +.98590611358921753738e+00;  P(305) = +.14092845069160408355e-01;

     % *****    Executable Statements    *******************************

     icheck=0;
215  delta=b-a;
     diff=0.5*delta;
     ip=1;
     jh=0;

220  % Apply 1-point Gauss formula (Midpoint rule).

     fncval=f(a+diff);
     % Don't write "0.5*(b+a)" above if the radix of arithmetic isn't 2.
     npts=1;
225  work(1)=fncval;
```

```
      acum=fncval*delta;
      result(1)=acum;

      for k = 2:8
230
      %   Go on to the next formula.

      pacum=acum;
      acum=0.0;
235
      %   Compute contribution to current estimate due to function
      %   values used in previous formulae.

      for kk = kx(k-1)+1 : kx(k)
240         for j = kl(kk) : kh(kk)
            acum=acum+(P(ip)*work(j));
            ip=ip+1;
          end;
        end;
245
      %   Compute contribution from new function values.

      jl=jh+1;
      jh=jl+jl-1;
250   j1=fl(k);
      j2=fh(k);
      for j = jl:jh
        x=P(ip)*diff;
        fncval=f(a+x)+f(b-x);
255     npts=npts+2;
        acum=acum+(P(ip+1)*fncval);
        if (j1 <= j2)
          work(j1)=fncval;
          j1=j1+1;
260     end
        ip=ip+2;
      end;
      acum=(diff)*acum+0.5*pacum;
      result(k)=acum;
265   err = abs(result(k)-result(k-1));
      int = result(k);
      if (err <= tol)
        return;
      end;
270   end; % for k=2:8
```

A.13   Ninomiya's 1980 Error Estimate

```
function [ int , err ] = est_ninomiya1980 ( f , a , b , tol )

    % the nodes
    xi = linspace(0,1,9)';
5   xi = a + (b - a) * sort( [ xi ; (xi(1)+xi(2))/2 ; (xi(end)+xi(end-1))/2 ] );

    % get the function values
    fx = f(xi);

10  % compute the quadrature and the difference
    err = 4736 * (b-a) / 2 / 468242775 * [ 3003 -16384 27720 -38220 56056 -64350 ...
        56056 -38220 27720 -16384 3003 ] * fx;
    int = (b-a) * [ 989 0 5888 -928 10496 -4540 10496 -928 5888 0 989 ] * fx / 28350 - err;
    err = abs(err);
15
end
```

A.14   Piessens *et al.* 's 1983 Error Estimate

```
    function [ int , err ] = est_piessens1983 ( f , a , b , tol )

       % the nodes
       xi = [ 0.9956571630258081e+00,    0.9739065285171717e+00, ...
5           0.9301574913557082e+00,    0.8650633666889845e+00, ...
            0.7808177265864169e+00,    0.6794095682990244e+00, ...
            0.5627571346686047e+00,    0.4333953941292472e+00, ...
            0.2943928627014602e+00,    0.1488743389816312e+00, ...
            0.0000000000000000e+00 ]';
10     xi = [ xi ; -xi(end-1:-1:1) ];

       % get the function values
       fx = f(a + (b-a) * (xi + 1)/2);

15     % the weights
       w = [ 0.6667134430868814e-01,    0.1494513491505806e+00, ...
            0.2190863625159820e+00,    0.2692667193099964e+00, ...
            0.2955242247147529e+00 ];
       w = [ w , w(end:-1:1) ] / 2;
20     wk = [ 0.1169463886737187e-01,   0.3255816230796473e-01, ...
            0.5475589657435200e-01,    0.7503967481091995e-01, ...
            0.9312545458369761e-01,    0.1093871588022976e+00, ...
            0.1234919762620659e+00,    0.1347092173114733e+00, ...
            0.1427759385770601e+00,    0.1477391049013385e+00, ...
25          0.1494455540029169e+00 ];
       wk = [ wk , wk(end-1:-1:1) ] / 2;

       % compute the quadratures
       g = (b-a) * w * fx(2:2:end-1);
30     gk = (b-a) * wk * fx;
       ik = (b-a) * wk * abs(fx - gk/(b-a));

       % compute the integral
       int = gk;

35     % compute the error estimate
       if abs(ik) > 0
           err = ik * min( 1 , (200 * abs(g - gk) / ik)^(3/2) );
       else
40         err = 0;
       end

    end
```

A.15   Laurie's 1983 Error Estimate

```
    function [ int , err ] = est_laurie1983 ( f , a , b , tol )
    %
    % Introduced "myeps" to avoid spurious failure due to
    % noise in the quadratures.
5
       % the nodes
       xi = [ 0.9956571630258081e+00,    0.9739065285171717e+00, ...
            0.9301574913557082e+00,    0.8650633666889845e+00, ...
            0.7808177265864169e+00,    0.6794095682990244e+00, ...
10          0.5627571346686047e+00,    0.4333953941292472e+00, ...
            0.2943928627014602e+00,    0.1488743389816312e+00, ...
            0.0000000000000000e+00 ]';
       xi = [ -xi ; xi(end-1:-1:1) ];

15     % get the function values
       fx = f(a + (b-a) * (xi + 1)/2);

       % the weights
       w = [ 0.6667134430868814e-01,    0.1494513491505806e+00, ...
20          0.2190863625159820e+00,    0.2692667193099964e+00, ...
            0.2955242247147529e+00 ];
       w = [ w , w(end:-1:1) ] / 2;
       wk = [ 0.1169463886737187e-01,   0.3255816230796473e-01, ...
            0.5475589657435200e-01,    0.7503967481091995e-01, ...
25          0.9312545458369761e-01,    0.1093871588022976e+00, ...
```

```
        0.1234919762620659e+00,     0.1347092173114733e+00, ...
        0.1427759385770601e+00,     0.1477391049013385e+00, ...
        0.1494455540029169e+00 ];
30   wk = [ wk , wk(end-1:-1:1) ] / 2;

     % compute the quadratures
     qb1 = (b-a) * w * fx(2:2:end-1);
     qa1 = (b-a) * wk * fx;

35   % get the double-nodes
     xi2 = [ (xi - 1)/2 ; (xi + 1)/2 ];
     fx2 = f(a + (b-a)*(xi2+1)/2);

     % compute the double quadratures
40   qb2 = (b-a) * [ w , w ] * fx2([ 2:2:20 , 23:2:41 ]') / 2;
     qa2 = (b-a) * [ wk , wk ] * fx2 / 2;

     % the integral
     int = qa2;

45   % test the conditions
     qab1 = qa1 - qb1; qaa = qa2 - qa1;
     qab2 = qa2 - qb2; qbb = qb2 - qb1;

50   % set some noise limits
     noise = 21 * eps * (b-a) * max(abs(fx));
     if abs(qab1) <= noise, qab1 = 0; end;
     if abs(qab2) <= noise, qab2 = 0; end;
     if abs(qaa) <= noise, qaa = 0; end;
55   if abs(qbb) <= noise, qbb = 0; end;

     if (qab1 == 0 || (0 <= qab2/qab1 && qab2/qab1 <= 1)) && ...
        (qaa == 0 || abs(qaa) <= abs(qbb))
        if qaa == 0
60          err = 0;
        else
            err = abs( qab2 * qaa / (qbb - qaa) );
        end;
     else
65        err = 2 * tol;
     end;

end
```

A.16   Kahaner and Stoer's 1983 Error Estimate

```
     function [ int , err ] = est_kahaner1983 ( f , a , b , tol )
     %
     %   Modified epsilon-algorithm slightly such that it abandons if
     %   two successive column entries are identical (or almost).
5
     % the nodes and function values
     n = 7;
     xi = a + (b-a) * linspace(0,1,2^(n-1) + 1)';
     fx = f(xi);
10
     % construct the basic quadratures
     for i=1:n
         q(i) = (b-a) * [ 1/2 ones(1,2^(i-1)-1) 1/2 ] * fx(1:2^(n-i):2^(n-1)+1) / 2^(i-1);
     end;
15
     % init the eps-table
     e = zeros(n);
     e(:,1) = q(:);
     for i=1:n-1
20       e(i,2) = 1.0 / (e(i+1,1) - e(i,1));
     end;
     for k=3:n
```

```
      for i=1:n-k+1
        d = e(i+1,k-1) - e(i,k-1);
25      if d == 0 || abs(2 * d / (e(i+1,k-1) + e(i,k-1))) < max(abs(fx))*eps*(b-a)*10
          int = e(i+1,k-1);
          if d == 0
            err = 0.0;
          else
30          err = tol * abs(d / (e(i+1,k-1) + e(i,k-1)));
          end;
          return;
        end;
        e(i,k) = e(i+1,k-2) + 1.0 / d;
35    end;
    end;

    % the integral is the last entry in the rightmost odd column
    k = mod(n,2);
40  int = e(2-k,n-1+k);
    int_prev1 = e(1+2*k,n-1-k);
    int_prev2 = e(4-k*2,n-3+k);
    err = abs(int - int_prev1) + abs(int - int_prev2);

45  end
```

A.17   Berntsen and Espelid's 1984 Error Estimate

```
function [ int , err ] = est_berntsen1984 ( f , a , b , tol )

    % the nodes
    xi21 = [-0.9937521706203895002602420359938, -0.9672268385663062943166222214908, ...
5       -0.9200993341504008287901871337715, -0.8533633645833172836472506385588, ...
        -0.7684399634756779086158778513065, -0.6671388041974123193059666699990, ...
        -0.5516188358872198070590187967247, -0.4243421202074387835736688885447, ...
        -0.2880213168024010966007925160655, -0.1455618541608950909370309823390, 0., ...
        0.1455618541608950909370309823390, 0.2880213168024010966007925160655, ...
10      0.4243421202074387835736688885447, 0.5516188358872198070590187967247, ...
        0.6671388041974123193059666699990, 0.7684399634756779086158778513065, ...
        0.8533633645833172836472506385588, 0.9200993341504008287901871337715, ...
        0.9672268385663062943166222214908, 0.9937521706203895002602420359938]';

15  % weights
    w21 = [0.0160172282577743333242187078818, 0.0369537897708524937998313495754, ...
        0.0571344254246857208283374330883, 0.0761001136283793020169811548244, ...
20      0.0934444234560338615532228314058, 0.1087972991671483776633502907833, ...
        0.1218314160537285341953229265800, 0.1322689386333374617811325750480, ...
        0.1398873947910731547220072730193, 0.1445244039899700590639482678913, ...
        0.1460811336496904271919052793570, 0.1445244039899700590639210750568, ...
        0.1398873947910731547220885567485, 0.1322689386333374617812315043383, ...
        0.1218314160537285341953052122553, 0.1087972991671483776634772784963, ...
25      0.0934444234560338615533044174843, 0.0761001136283793020170415377821, ...
        0.0571344254246857208283635545266, 0.0369537897708524937999293341024, ...
        0.0160172282577743333242239053239];
    w20 = [0.0106185068786295819657339016891, 0.0556095497747986766746793, ...
        0.0213509066165672182894520098703, 0.1310657484884345603353592700481, ...
30      0.0186790125246841257874886375194, 0.2027098392220866781419074314089, ...
        0.0105576172897147219174700745, 0.2581369541881641159743375270903, ...
        0.0029943519231780123242390972415, 0.2882775126546652832691175490943, ...
        0.2882775126546652832691076751603, 0.0029943519231780123242290329114, ...
        0.2581369541881641159744166637859, 0.0105576172897147219316248009893, ...
35      0.2027098392220866781418443273205, 0.0186790125246841257874671063730, ...
        0.1310657484884345603353692369995, 0.0213509066165672182893290880775, ...
        0.0556095497747986766748067438711, 0.0106185068786295819657745168233];

    % compute stuff
    fx = f(xi21);
40  g21 = (b-a) * w21 * fx / sum(w21);
    q20 = (b-a) * w20 * fx([1:10,12:21]) / sum(w20);
```

```
        % the integral
        int = g21;
45
        % the error
        err = abs( g21 - q20 );

    end
```

A.18 Berntsen and Espelid's 1991 Error Estimate

```
    function [ int , err ] = est_berntsen1991 ( f , a , b , tol )

        % some constants
        r_crit = 1/4;
5       alpha = (21 - (2*10 - 3)) / 4;
        K = 3;

        % the nodes and function values
        xi = -cos( [0:20] / 20 * pi )';
10      fx = f(a + (b-a) * (xi + 1) / 2);

        % legendre base...
        N = [ ones(21,1) , xi ];
        for i=3:21, N(:,i) = 2*xi.*N(:,i-1)-N(:,i-2); end;
15      [ N , r ] = qr(N);
        N = N';

        w = [.25062656641604010e-2, .23978647239640935e-1, .48754129106911638e-1, ...
20           .71202738820962617e-1, .92397630395614637e-1, .11102364439825740, ...
             .12711755339015786, .13992786619265645, .14941903247367981, ...
             .15512023618056280, .15710451227479091, .15512023618056280, ...
             .14941903247367981, .13992786619265645, .12711755339015786, ...
             .11102364439825740, .92397630395614637e-1, .71202738820962617e-1, ...
             .48754129106911638e-1, .23978647239640935e-1, .25062656641604010e-2];
25      N = N * norm(w);

        % compute the coefficients
        int = (b-a) * w * fx / sum(w);
        e = N * fx;
30      E = sqrt(e(2:2:20).^2 + e(3:2:21).^2);
        r = E(end-K:end-1) ./ max(E(end-K+1:end),eps);
        r_max = max(r);

        % so many decisions...
35      if r_max > 1
            err = max(E(end-K:end));
        elseif r_max > r_crit;
            err = 10 * r_max * E(end);
        else
40          err = 10 * r_crit^(1-alpha) * r_max^alpha * E(end);
        end

    end
```

A.19 Favati *et al.*'s 1991 Error Estimate

```
    function [ int , err ] = est_favati1991 ( f , a , b , tol )

        % nodes and weights
        xi15 = [ -0.99145537112081263 -0.94910791234275852 -0.86486442335976907 ...
5            -0.74153118559939443 -0.58608723546769113 -0.40584515137739716 -0.20778495500789846 ...
             0.00000000000000000 0.20778495500789846 0.40584515137739716 0.58608723546769113 ...
             0.74153118559939443 0.86486442335976907 0.94910791234275852 0.99145537112081263 ]';
        w7 = [ 0.12948496616886969 0.27970539148927666 0.38183005050511894 ...
```

```
            0.41795918367346938 0.38183005050511894 0.27970539148927666 0.12948496616886969 ];
10     w15 = [  0.02293532201052922 0.06309209262997855 0.10479001032225018 ...
            0.14065325971552591 0.16900472663926790 0.19035057806478540 0.20443294007529889 ...
            0.20948214108472782 0.20443294007529889 0.19035057806478540 0.16900472663926790 ...
            0.14065325971552591 0.10479001032225018 0.06309209262997855 0.02293532201052922 ];
       xi21 = [  -0.99565716302580808 ...
15          -0.97390652851717172 -0.93015749135570822 -0.86506336668898451 -0.78081772658641689 ...
            -0.67940956829902440 -0.56275713466860468 -0.43339539412924719 -0.29439286270146019 ...
            -0.14887433898163121 0.00000000000000000 0.14887433898163121 0.29439286270146019 ...
            0.43339539412924719 0.56275713466860468 0.67940956829902440 0.78081772658641689 ...
            0.86506336668898451 0.93015749135570822 0.97390652851717172 0.99565716302580808 ]';
20     w10 = [ 0.06667134430868813 0.14945134915058059 ...
            0.21908636251598204 0.26926671930999635 0.29552422471475287 0.29552422471475287 ...
            0.26926671930999635 0.21908636251598204 0.14945134915058059 0.06667134430868813 ];
       w21 = [ 0.01169463886737187 ...
            0.03255816230796472 0.05475589657435199 0.07503967481091995 0.09312545458369760 ...
25          0.10938715880229764 0.12349197626206585 0.13470921731147332 0.14277593857706008 ...
            0.14773910490133849 0.14944555400291690 0.14773910490133849 0.14277593857706008 ...
            0.13470921731147332 0.12349197626206585 0.10938715880229764 0.09312545458369760 ...
            0.07503967481091995 0.05475589657435199 0.03255816230796472 0.01169463886737187 ];

30     % get the function values
       fx15 = f(a + (b-a) * (xi15 + 1)/2);
       fx21 = f(a + (b-a) * (xi21 + 1)/2);

       % compute the four quadratures
35     q7  = (b-a) * w7  * fx15(2:2:14) / sum(w7);
       q10 = (b-a) * w10 * fx21(2:2:20) / sum(w10);
       q15 = (b-a) * w15 * fx15 / sum(w15);
       q21 = (b-a) * w21 * fx21 / sum(w21);

40     % if the difference between the last two estimates is noise,
       % just split.
       if abs(q21-q15) < 100 * (b-a) * eps * max(abs(fx21))
          int = q21;
          err = abs(q21-q15);
45        return;
       end;

       % compute the bounds
       b1 = (q7 - q21) / (q15 - q21);
50     b2 = (q10 - q21) / (q15 - q21);
       if 1 - b1 + b2 == 0, s0 = inf;
       else s0 = b2 / ( 1 - b1 + b2 );
       end;
       delta = (1 + b1 + b2)^2 - 8*b2;
55     s1 = (1 + b1 + b2 - sqrt(delta)) / 4;
       s2 = (1 + b1 + b2 + sqrt(delta)) / 4;
       if b1 > 1 + b2 && b2 > 0
          alpha = s0; beta = min([ b2/2 , 1/2 , s1 ]);
       elseif b1 > 0 && b1 <= 1 + b2 && b2 < 0
60        alpha = max(b2/2,s1); beta = min(b1/2,1/2);
       elseif b1 > 0 && b1 > 1 + b2 && b2 < 0
          alpha = max(b2/2,s1); beta = min([1/2,1/2,s0]);
       elseif b1 < 0 && b1 < 1 + b2 && b2 < 0
          alpha = max([b1/2,b2/2,s0]); beta = min(1/2,s2);
65     elseif b1 < 0 && b1 >= 1 + b2 && b2 < 0
          alpha = max(b1/2,b2/2); beta = min(1/2,s2);
       elseif b1 < 0 && b2 > 0
          alpha = b1/2; beta = min([1/2,b2/2,s0]);
       else
70        int = q21;
          err = 2*tol;
          return;
       end;

75     % compute the error bound an the integral
       int = q21;
       err = max( abs(alpha) , abs(beta) ) * abs(q15 - q21);

   end
```

A.20   Gander and Gautschi's 2001 Error Estimate

```
function [ int , err ] = est_gander2001 ( f , a , b , tol )

      % the nodes and function values
      c = (a + b)/2;
 5    h = (b - a)/2;
      alpha = sqrt(2/3);
      beta = 1/sqrt(5);
      xi = a + (b-a)*([-1 c-alpha*h c-beta*h c c+beta*h c+alpha*h 1]' + 1) / 2;
      fx = f(xi);
10
      % compute the rules
      gl = (b-a) * [1 5 5 1] * fx(1:2:7) / 12;
      gk = (b-a) * [77 432 625 672 625 432 77] * fx / 2940;

15    % the int and the error
      int = gk;
      err = abs( gl - gk );

end
```

A.21   Venter and Laurie's 2002 Error Estimate

```
function [ int , err ] = est_venter2002 ( f , a , b , tol )

      % the nodes and weigths
      xi = [ 0.0 0.816496580927260 0.4744811695891284 0.9744079242540920 ...
 5        0.2467208395344672 0.6676815505186759 0.9176634454676027 0.9967032921348130 ...
          0.1245872428182431 0.3640598996653803 0.5761686879241989 0.7479928493493064 ...
          0.8729905887501410 0.9511171199458969 0.9890095712955333 0.9995868407682468 ]';
      w = [ 0.15625e-1/2 0.78125e-2 0.1329095288597234e-1 0.2334047114027659e-2 ...
10        0.1501395420099286e-1 0.1075598273807455e-1 0.4865753315451455e-2 0.6143097454811318e-3 ...
          0.1547045578486217e-1 0.1427638731917522e-1 0.1210082672197449e-1 0.9309261194540840e-2 ...
          0.6314881366348385e-2 0.3519722515847501e-2 0.1353274640802211e-2 0.1551904564491656e-3 ]';

      % get the function values
15    fx = [ f(a + (b-a) * (xi + 1)/2) f(a + (b-a) * (-xi + 1)/2) ]';

      % compute the quadrature rules
      for i=1:5
          q(i) = (b-a) * 2^(8-i) * sum( fx(:,1:2^(i-1)) * w(1:2^(i-1)) ) / 2;
      end;
20
      % compute the error estimates
      E = max( abs( q(2:end) - q(1:end-1) ) , 2.2e-16 );
      hint = E(2:end) ./ E(1:end-1);
25    eps = hint .* E(2:end);

      % loop...
      for i=1:length(eps)
          if hint(i) > 0.1
30            err = 2*tol;
              int = q(i+2);
              % disp('bail hint...');
              return;
          elseif eps(i) < tol
35            err = eps(i);
              int = q(i+2);
              % disp('bail conv...');
              return;
          end;
      end;
```

```
40
      % ran out of quadratures
      err = 2*tol;
      int = q(end);

45  end
```

A.22 The New Trivial Error Estimate

```
function [ int , err ] = est_gonnet2008b ( f , a , b , tol )

      % the nodes and function values
5     n = 10;
      xi_1 = -cos([0:n]/n*pi)';
      xi_2 = -cos([0:2*n]/2/n*pi)';
      fx = f(a + (b-a)*(xi_2 + 1)/2);

10    % compute the coefficients
      V_1 = [ ones(size(xi_1)) xi_1 ];
      for i=3:n+1
          V_1 = [ V_1 , ((2*i-3) / (i-1) * xi_1 .* V_1(:,i-1) - (i-2) / (i-1) * V_1(:,i-2)) ];
      end;
15    V_2 = [ ones(size(xi_2)) xi_2 ];
      for i=3:2*n+1
          V_2 = [ V_2 , ((2*i-3) / (i-1) * xi_2 .* V_2(:,i-1) - (i-2) / (i-1) * V_2(:,i-2)) ];
      end;
      for i=1:n+1, V_1(:,i) = V_1(:,i) * sqrt(4*i-2)/2; end;
20    for i=1:2*n+1, V_2(:,i) = V_2(:,i) * sqrt(4*i-2)/2; end;
      c_1 = V_1 \ fx([1:2:2*n+1]);
      c_2 = V_2 \ fx;

      % compute the integral
25    w = [ 1/sqrt(2) , zeros(1,2*n) ];
      int = (b-a) * w * c_2;

      % compute the errors left and right
      scale = 0.0009745855668952330808842;
30    c_diff = norm([c_1;zeros(n,1)]-c_2);
      err = (b-a) * c_diff;

end
```

A.23 The New Refined Error Estimate

```
function [ int , err ] = est_gonnet2008 ( f , a , b , tol )

      % the nodes and function values
5     n = 10;
      xi = -cos([0:n]/n*pi)';
      xil = (xi - 1)/2; xir = (xi + 1)/2;
      fx = f(a + (b-a)*(xi + 1)/2);
      fxl = f(a + (b-a)*(xil+1)/2);
10    fxr = f(a + (b-a)*(xir+1)/2);

      % compute the coefficients
      V = [ ones(size(xi)) xi ];
      Vl = [ ones(size(xi)) xil ]; Vr = [ ones(size(xi)) xir ];
15    for i=3:n+2
          V = [ V , ((2*i-3) / (i-1) * xi .* V(:,i-1) - (i-2) / (i-1) * V(:,i-2)) ];
          Vl = [ Vl , ((2*i-3) / (i-1) * xil .* Vl(:,i-1) - (i-2) / (i-1) * Vl(:,i-2)) ];
          Vr = [ Vr , ((2*i-3) / (i-1) * xir .* Vr(:,i-1) - (i-2) / (i-1) * Vr(:,i-2)) ];
      end;
20    for i=1:n+2
          V(:,i) = V(:,i) * sqrt(4*i-2)/2;
          Vl(:,i) = Vl(:,i) * sqrt(4*i-2)/2;
```

```
         Vr(:,i) = Vr(:,i) * sqrt(4*i-2)/2;
       end;
25     c_old = V(:,1:n+1) \ fx;
       c_left = V(:,1:n+1) \ fxl;
       c_right = V(:,1:n+1) \ fxr;

       % shift matrix
30     T_left = V(:,1:n+1) \ Vl(:,1:n+1);
       T_right = V(:,1:n+1) \ Vr(:,1:n+1);

       % compute the integral
       w = [ 1/sqrt(2) , zeros(1,n) ];
35     int = (b-a) * w * (c_left + c_right) / 2;

       % compute the errors left and right
       b_old = [0., .10620845052443988729141015545e-4, 0., ...
         .2271305231870350977068612042966e-4, 0., .5818880429155495585524788754554e-4, ...
40       0., .3346672959803132367207928892106e-3, 0., ...
         -.130155910942313246816263138278e-2, 0., ...
         .856102717782094742857796872476e-3]';
       b_left = 2^(n+1) * [.2762135864009951267190798289716e-3, ...
         .499657655549222744976243907697e-3, .523434680236984721004073014910e-3, ...
45       .257874261861280205575039917099e-3, -.748342903416082700961183312797e-3, ...
         -.917669983469695220709693581407e-3, .549500980250147994974675357971e-3, ...
         .300060029460695461688333929627e-3, -.276678492776836119682157766061e-3, ...
         .761043473405371645812866738709e-4, -.918691057168920637420186132198e-5, ...
         .418018905167038448661033629139e-6]';
50     b_right = 2^(n+1) * [-.2762135864009951267190798289716e-3, ...
         .499657655549222744976243907697e-3, -.523434680236984721004073014910e-3, ...
         .257874261861280205575039917099e-3, .748342903416082700961183312797e-3, ...
         -.917669983469695220709693581407e-3, -.549500980250147994974675357971e-3, ...
55       .300060029460695461688333929627e-3, .276678492776836119682157766061e-3, ...
         .761043473405371645812866738709e-4, .918691057168920637420186132198e-5, ...
         .418018905167038448661033629139e-6]';
       scale = norm(b_old) / norm((b_old - b_left));
       fn_left = norm(c_left - T_left*c_old) / norm((b_old - b_left));
       fn_right = norm(c_right - T_right*c_old) / norm((b_old - b_right));
60     err_left = (b-a)/2 * fn_left * norm(b_old);
       if max(abs(V(:,1:n+1) * T_left * c_old - fxl)(2:end-1) ...
         1.1 * fn_left * abs( V * b_left )(2:end-1)) > cond(V)*eps*norm(c_old)
         err_left = (b-a)/2 * norm(c_left - T_left*c_old);
       else
65       err_left = (b-a)/2 * fn_left * norm(b_old);
       end;
       if max(abs(V(:,1:n+1) * T_right * c_old - fxr)(2:end-1) - ...
         1.1 * fn_right * abs( V * b_right )(2:end-1)) > cond(V)*eps*norm(c_old)
         err_right = (b-a)/2 * norm(c_right - T_right*c_old);
70     else
         err_right = (b-a)/2 * fn_right * norm(b_old);
       end;
       err = err_left + err_right;

75     end
```

Appendix B

Algorithms in Matlab

B.1 An Adaptive Quadrature Algorithm using Interpolation Updates

```
     function [ int , err , nr_points ] = int_gonnet2008d ( f , a , b , tol )
     %INT_GONNET2008D  evaluates an integral using adaptive quadrature. The
     %    algorithm uses a 9-point RMS quadrature rule as a basic stencil
     %    and extends an interpolation of the integrand generated in the
   5 %    previous level of recursion by adding the new nodes in the interval.
     %    If a maximum number of nodes has been reached or the change in the
     %    interpolation is too large, the interpolation is reset. The error
     %    estimate is computed from the L2-norm of the difference between two
     %    successive interpolations.
  10 %
     %    INT = INT_GONNET2008D ( F , A , B , TOL ) approximates the integral
     %    of F in the interval [A,B] up to the relative tolerance TOL. The
     %    integrand F should accept a vector argument and return a vector
     %    result containing the integrand evaluated at each element of the
  15 %    argument.
     %
     %    [INT,ERR,NR_POINTS] = INT_GONNET2008D ( F , A , B , TOL ) returns
     %    ERR, an estimate of the absolute integration error as well as
     %    NR_POINTS, the number of function values for which the integrand
  20 %    was evaluated. The value of ERR may be larger than the requested
     %    tolerance, indicating that the integration may have failed.
     %
     %    INT_GONNET2008D halts with a warning if the integral is or appears
     %    to be divergent.
  25 %
     %    Reference: "Adaptive Quadrature Re-Revisited", P. Gonnet, PhD Thesis,
     %        submitted 2008.

     % declare persistent variables
  30 persistent n_init n_max xi b_init V Vx xx Vinv Vcond T_left T_right w ...
         alpha beta gamma T_incr xi_left xi_right sel_left sel_right

     % have the persistent variables been declared already?
     if ~exist('w') || isempty(w)

  35     % the nodes and function values
```

201

```
     n_init = 8;
     n_max = 20;
     % xi = linspace(-1,1,n_init+1)';
40   xi = [ -1, -7/8, -3/4, -1/2, 0, 1/2, 3/4, 7/8, 1 ]';
     xi_left = [ 2 6 7 8 ]; sel_left = [ 1 1 2 3 4 4 4 4 5 ];
     xi_right = [ 2 3 4 8 ]; sel_right = [ 5 5 5 5 6 7 8 8 9 ];

     % the initial vector b
45   b_init = [ 0., 0.0016431884229206290082, 0., -0.0053032949213072811071, 0., ...
     0.0062061518050860500735, 0., -0.0062710590362925230324, 0., ...
     0.0034165926622357227290 , zeros(1,n_max-n_init) ]';

     % compute the vandermonde-like matrix
50   V = [ ones(size(xi)) xi ];
     xx = linspace(-1,1,200)'; Vx = [ ones(size(xx)) xx ];
     for i=3:n_max+2
        V = [ V , ((2*i-3) / (i-1) * xi .* V(:,i-1) - (i-2) / (i-1) * V(:,i-2)) ];
        Vx = [ Vx , ((2*i-3) / (i-1) * xx .* Vx(:,i-1) - (i-2) / (i-1) * Vx(:,i-2)) ];
55   end;
     for i=1:n_max+2
        V(:,i) = V(:,i) * sqrt(4*i-2)/2;
        Vx(:,i) = Vx(:,i) * sqrt(4*i-2)/2;
     end;
60   Vinv = inv(V(:,1:n_init+1));
     Vcond = cond(V);

     % shift matrices
     T_left = [ ... ];
65   T_right = [ ... ];
     T_left = T_left(1:n_max+2,1:n_max+2);
     T_right = T_right(1:n_max+2,1:n_max+2);

     % the coefficient matrix (not explicit)
70   k = [0:n_max+1]';
     alpha = sqrt( (k+1).^2 ./ (2*k+1) ./ (2*k+3) );
     beta = 0 * k;
     gamma = [ sqrt(2) ; sqrt( k(2:end).^2 ./ (4*k(2:end).^2 - 1) ) ];
     T_incr = diag(alpha(1:end-1),-1) - diag(beta) + diag(gamma(2:end),1);
75
     % compute the integral
     w = [ sqrt(2) , zeros(1,n_max) ] / 2; % legendre

     end; % if exist('w')
80
     % create the original datatype
     ivals = struct( ...
        'a', [], 'b',[], 'n', [], 'n_old', [], ...
        'c', [], 'c_old', [], ...
85      'b_new', [], 'b_old', [], ...
        'fx', [], ...
        'int', [], ...
        'err', [], ...
        'tol', [], ...
90      'depth', [], ...
        'ndiv' , [] );

     % compute the first interval
     points = a + (b-a) * (xi + 1) / 2;
95   fx = f(points); nans = [];
     for i=1:length(fx), if ~isfinite(fx(i)), nans = [ i , nans ]; fx(i) = 0.0; end; end;
     c_new = [ Vinv * fx ; zeros(n_max-n_init,1) ];
     b_new = b_init; n = n_init;
     for i=1:length(nans)
100     b_new(1:end-1) = (T_incr(2:end,1:end-1) - diag(ones(n_max,1)*xi(nans(i)),1)) \ b_new(2:end);
        b_new(end) = 0; c_new = c_new - c_new(n+1) / b_new(n+1) * b_new(1:end-1);
        n = n - 1;
     end;
     ivals(1).fx = fx;
105  ivals(1).c = c_new;
     ivals(1).c_old = c_new;
     ivals(1).a = a; ivals(1).b = b; ivals(1).n = n_init;
     ivals(1).int = (b-a) * w * ivals(1).c;
     ivals(1).b_new = b_new;
110  ivals(1).err = abs((b-a) * norm(ivals(1).c));
     ivals(1).tol = tol;
```

```
       ivals(1).depth = 1;
       ivals(1).ndiv = 0;

115    % init some globals
       int = ivals(1).int; err = ivals(1).err; nr_ivals = 1;
       int_final = 0; err_final = 0; err_excess = 0;
       i_max = 1; nr_points = n_init+1;
       ndiv_max = 20; theta_1 = 0.1;
120
       % main loop
       while true

           % compute the left interval
125        a_left = ivals(i_max).a; b_left = (ivals(i_max).a + ivals(i_max).b)/2;
           h_left = b_left - a_left;
           points_left = a_left + h_left * (xi + 1) / 2;
           fx_left = ivals(i_max).fx(sel_left);
           fx_left(xi_left) = f(points_left(xi_left));
130        n_left = ivals(i_max).n;
           b_old = T_left * ivals(i_max).b_new;
           c_old = T_left(1:n_max+1,1:n_max+1) * ivals(i_max).c;
           nr_points = nr_points + n_init/2;
           if (n_left + n_init/2 <= n_max)
135            b_new = b_old; c_left = c_old;
               for i=xi_left
                   if ~isfinite(fx_left(i)), continue; end;
                   g_k = V(i,1:n_left+1) * c_left(1:n_left+1);
                   p_k = V(i,1:n_left+2) * b_new(1:n_left+2);
140                a_k = (fx_left(i) - g_k) / p_k;
                   n_left = n_left + 1;
                   c_left = c_left + a_k * b_new(1:end-1);
                   b_new = (T_incr - diag(xi(i)*ones(n_max+2,1))) * b_new;

145            end;
               nc = norm(c_left);
           end;
           if n_left == ivals(i_max).n || ( nc > 0 && norm(c_left - c_old) / nc > theta_1 )

150            nans = [];
               for i=1:length(fx_left), if ~isfinite(fx_left(i)), fx_left(i) = 0.0; nans = [ i , nans ]; end; end;
               c_left = [ Vinv * fx_left ; zeros(n_max-n_init,1) ];
               b_new = b_init;
               n_left = n_init;
155            for i=nans
                   b_new(1:end-1) = (T_incr(2:end,1:end-1) - diag(ones(n_max,1)*xi(i),1)) \ b_new(2:end);
                   b_new(end) = 0; c_left = c_left - c_left(n_left+1) / b_new(n_left+1) * b_new(1:end-1);
                   n_left = n_left - 1;
                   fx_left(i) = NaN;
160            end;
               end;
           int_left = h_left * w(1:n_left+1) * c_left(1:n_left+1);
           err_left = h_left * norm(c_left - c_old);
           if abs(ivals(i_max).int) > 0 && ...
165            int_left / ivals(i_max).int > 1
               ndiv_left = ivals(i_max).ndiv + 1;
           else
               ndiv_left = ivals(i_max).ndiv;
           end;
170        if ndiv_left > ndiv_max && 2*ndiv_right > ivals(i_max).depth
               warning(sprintf('possibly_divergent_integral_in_the_interval_[%e,%e]!',a_left,b_left));
               int = sign(int) * Inf;
               return;
           end;
175        if points_left(end) <= points_left(end-1) || ...
               err_left < abs(int_left) * Vcond * eps
               err_excess = err_excess + err_left;
               int_final = int_final + int_left;

180        else
               nr_ivals = nr_ivals + 1;
               ivals(nr_ivals).c = c_left;
               ivals(nr_ivals).c_old = c_old;
               ivals(nr_ivals).a = a_left; ivals(nr_ivals).b = b_left;
185            ivals(nr_ivals).fx = fx_left;
               ivals(nr_ivals).b_new = b_new;
```

```
            ivals(nr_ivals).b_old = b_old;
            ivals(nr_ivals).n = n_left;
            ivals(nr_ivals).n_old = ivals(i_max).n;
190         ivals(nr_ivals).int = int_left;
            ivals(nr_ivals).err = err_left;
            ivals(nr_ivals).tol = ivals(i_max).tol / sqrt(2);
            ivals(nr_ivals).depth = ivals(i_max).depth + 1;
            ivals(nr_ivals).ndiv = ndiv_left;
195     end;


        % compute the right interval
        a_right = (ivals(i_max).a + ivals(i_max).b)/2; b_right = ivals(i_max).b;
200     h_right = b_right - a_right;
        points_right = a_right + h_right * (xi + 1) / 2;
        fx_right = ivals(i_max).fx(sel_right);
        fx_right(xi_right) = f(points_right(xi_right));
        nr_points = nr_points + n_init/2;
205     n_right = ivals(i_max).n;
        b_old = T_right * ivals(i_max).b_new;
        c_old = T_right(1:n_max+1,1:n_max+1) * ivals(i_max).c;
        if (n_right + n_init/2 <= n_max)
            b_new = b_old; c_right = c_old;
210         for i=xi_right
                if ~isfinite(fx_right(i)), continue; end;
                g_k = V(i,1:n_right+1) * c_right(1:n_right+1);
                p_k = V(i,1:n_right+2) * b_new(1:n_right+2);
                a_k = (fx_right(i) - g_k) / p_k;
215             n_right = n_right + 1;
                c_right = c_right + a_k * b_new(1:end-1);
                b_new = (T_incr - diag(xi(i)*ones(n_max+2,1))) * b_new;
            end;
            nc = norm(c_right);
220     end;
        if n_right == ivals(i_max).n || ( nc > 0 && norm(c_right - c_old) / nc > theta_1 )
            fx = f(points_right); nans = [];
            for i=1:length(fx), if ~isfinite(fx(i)), fx(i) = 0.0; nans = [ i , nans ]; end; end;
            c_right = [ Vinv * fx ; zeros(n_max-n_init,1) ];
225         b_new = b_init;
            n_right = n_init;
            for i=nans
                b_new(1:end-1) = (T_incr(2:end,1:end-1) - diag(ones(n_max,1)*xi(i),1)) \ b_new(2:end);
                b_new(end) = 0; c_right = c_right - c_right(n_right+1) / b_new(n_right+1) * b_new(1:end-1);
230             n_right = n_right - 1;
                fx_right(i) = NaN;
            end;
        end;
        int_right = h_right * w(1:n_right+1) * c_right(1:n_right+1);
235     err_right = h_right * norm(c_right - c_old);
        if abs(ivals(i_max).int) > 0 && ...
            int_right / ivals(i_max).int > 1
            ndiv_right = ivals(i_max).ndiv + 1;
        else
240         ndiv_right = ivals(i_max).ndiv;
        end;
        if ndiv_right > ndiv_max && 2*ndiv_right > ivals(i_max).depth
            warning(sprintf('possibly␣divergent␣integral␣in␣the␣interval␣[%e,%e]!',a_right,b_right));
            int = sign(int) * Inf;
245         return;
        end;
        if points_right(end) <= points_right(end-1) || ...
            err_right < abs(int_right) * Vcond * eps
            err_excess = err_excess + err_right;
250         int_final = int_final + int_right;
        else
            nr_ivals = nr_ivals + 1;
            ivals(nr_ivals).c = c_right;
            ivals(nr_ivals).c_old = c_old;
255         ivals(nr_ivals).a = a_right; ivals(nr_ivals).b = b_right;
            ivals(nr_ivals).fx = fx_right;
            ivals(nr_ivals).b_new = b_new;
            ivals(nr_ivals).b_old = b_old;
            ivals(nr_ivals).int = int_right;
260         ivals(nr_ivals).n = n_right;
            ivals(nr_ivals).n_old = ivals(i_max).n;
```

```
          ivals(nr_ivals).err = err_right;
          ivals(nr_ivals).tol = ivals(i_max).tol / sqrt(2);
          ivals(nr_ivals).depth = ivals(i_max).depth + 1;
265       ivals(nr_ivals).ndiv = ndiv_right;
          end;

          % copy last interval into i_max
          ivals(i_max) = ivals(nr_ivals);
270       nr_ivals = nr_ivals - 1;

          % compute the running err and new max
          i_max = 1; i_min = 1; err = err_final; int = int_final;
          for i=1:nr_ivals
275           if ivals(i).err > ivals(i_max).err, i_max = i;
              elseif ivals(i).err < ivals(i_min).err, i_min = i; end;
              err = err + ivals(i).err;
              int = int + ivals(i).int;
          end;
280
          % nuke smallest element if stack is larger than 500
          if nr_ivals > 500
              err_final = err_final + ivals(i_min).err;
              int_final = int_final + ivals(i_min).int;
285           ivals(i_min) = ivals(nr_ivals);
              if i_max == nr_ivals, i_max = i_min; end;
              nr_ivals = nr_ivals - 1;
          end;

290       % are we there yet?
          if err < abs(int) * tol || ...
              (err_final > abs(int) * tol && err - err_final < abs(int) * tol) || ...
              nr_ivals == 0
              break;
295       end;

          end; % main loop

          % clean-up and tabulate
300       err = err + err_excess;

      end
```

B.2   A Doubly-Adaptive Quadrature Algorithm using the Naive Error Estimate

```
      function [ int , err , nr_points ] = int_gonnet2008c ( f , a , b , tol )
      %INT_GONNET2008C  evaluates an integral using adaptive quadrature. The
      %   algorithm uses Clenshaw-Curtis quadrature rules of increasing
      %   degree in each interval and bisects the interval if either the
5     %   function does not appear to be smooth or a rule of maximum
      %   degree has been reached. The error estimate is computed from the
      %   L2-norm of the difference between two successive interpolations
      %   of the integrand over the nodes of the respective quadrature rules.
      %
10    %   INT = INT_GONNET2008C ( F , A , B , TOL ) approximates the integral
      %   of F in the interval [A,B] up to the relative tolerance TOL. The
      %   integrand F should accept a vector argument and return a vector
      %   result containing the integrand evaluated at each element of the
      %   argument.
15    %
      %   [INT,ERR,NR_POINTS] = INT_GONNET2008C ( F , A , B , TOL ) returns
      %   ERR, an estimate of the absolute integration error as well as
      %   NR_POINTS, the number of function values for which the integrand
      %   was evaluated. The value of ERR may be larger than the requested
20    %   tolerance, indicating that the integration may have failed.
      %
```

```
   %   INT_GONNET2008C halts with a warning if the integral is or appears
   %   to be divergent.
   %
25 %   Reference: "Adaptive Quadrature Re-Revisited", P. Gonnet, submitted
   %      to ACM Transactions on Mathematical Software, 2008.

   % declare persistent variables
   persistent n xi_1 xi_2 xi_3 xi_4 b_1_def b_2_def b_3_def b_4_def ...
30      V_1 V_2 V_3 V_4 V_1_inv V_2_inv V_3_inv V_4_inv xx Vcond T_left T_right w U

   % have the persistent variables been declared already?
   if ~exist('U') || isempty(U)

35     % the nodes and newton polynomials
       n = [4,8,16,32];
       xi_1 = -cos([0:n(1)]/n(1)*pi)';
       xi_2 = -cos([0:n(2)]/n(2)*pi)';
       xi_3 = -cos([0:n(3)]/n(3)*pi)';
40     xi_4 = -cos([0:n(4)]/n(4)*pi)';
       b_1_def = [0., .23328473740792172363783657854e-1, 0., ...
           -.831479419283098085685277496071e-1, 0.,0.0541462136776153483932540272848 ]';
       b_2_def = [0., .883654308363339862264532494396e-4, 0., ...
           .238811521522368331303214066075e-3, 0., 0.,0.0034165926622357227892690737979 ]';
45     0., -.520710690438660595086839959882e-2, 0.,0.0034165926622357227892690737979 ]';
       b_3_def = [0., .379785635776247894184454273159e-7, 0., ...
           .655473977795402040043020497901e-7, 0., .103479954638984842226816620692e-6, ...
           0., .173700624961660596894381303819e-6, 0., .337719613424065357737369982062e-6, ...
           0., .877423283560614343733565759649e-6, 0., .515657204371051131603503471028e-5, ...
50     0.,-.203244736027387801432055290742e-4, 0.,0.0000134265158311651777460545854542 ]';
       b_4_def = [0., .703046511513775683031092069125e-13, 0., ...
           .110617117381148770138566741591e-12, 0., .146334657087392356024202217074e-12, ...
           0., .184948444492681259461791759092e-12, 0., .231429962470609662207589449428e-12, ...
55     0., .291520062115989014852816512412e-12, 0., .373653379768759953435020783965e-12, ...
           0., .491840406039799844946199367185e-12, 0., .671514395653454630785723660045e-12, ...
           0., .963162916392726862525650710866e-12, 0., .147853378943890691325323722031e-11, ...
           0., .250420145651013003355273649380e-11, 0., .495516257435784759806147914867e-11, ...
           0., .130927034717608716480686641267e-10, 0., .779528640561654357271364483150e-10, ...
60     0., -.309866395328070487426324760520e-9, 0., .205573220292667201732878773151e-9]';

       % compute the coefficients
       V_1 = [ ones(size(xi_1)) xi_1 ];
       V_2 = [ ones(size(xi_2)) xi_2 ];
65     V_3 = [ ones(size(xi_3)) xi_3 ];
       V_4 = [ ones(size(xi_4)) xi_4 ];
       xil = (xi_4 - 1) / 2; xir = (xi_4 + 1) / 2;
       Vl = [ ones(size(xil)) xil ]; Vr = [ ones(size(xir)) xir ];
       xx = linspace(-1,1,500)'; Vx = [ ones(size(xx)) xx ];
       for i=3:n(1)+1
70         V_1 = [ V_1 , ((2*i-3) / (i-1) * xi_1 .* V_1(:,i-1) - (i-2) / (i-1) * V_1(:,i-2)) ];
       end;
       for i=3:n(2)+1
           V_2 = [ V_2 , ((2*i-3) / (i-1) * xi_2 .* V_2(:,i-1) - (i-2) / (i-1) * V_2(:,i-2)) ];
75     end;
       for i=3:n(3)+1
           V_3 = [ V_3 , ((2*i-3) / (i-1) * xi_3 .* V_3(:,i-1) - (i-2) / (i-1) * V_3(:,i-2)) ];
       end;
       for i=3:n(4)+1
           V_4 = [ V_4 , ((2*i-3) / (i-1) * xi_4 .* V_4(:,i-1) - (i-2) / (i-1) * V_4(:,i-2)) ];
80         Vx = [ Vx , ((2*i-3) / (i-1) * xx .* Vx(:,i-1) - (i-2) / (i-1) * Vx(:,i-2)) ];
           Vl = [ Vl , ((2*i-3) / (i-1) * xil .* Vl(:,i-1) - (i-2) / (i-1) * Vl(:,i-2)) ];
           Vr = [ Vr , ((2*i-3) / (i-1) * xir .* Vr(:,i-1) - (i-2) / (i-1) * Vr(:,i-2)) ];
       end;
       for i=1:n(1)+1, V_1(:,i) = V_1(:,i) * sqrt(4*i-2)/2; end;
85     for i=1:n(2)+1, V_2(:,i) = V_2(:,i) * sqrt(4*i-2)/2; end;
       for i=1:n(3)+1, V_3(:,i) = V_3(:,i) * sqrt(4*i-2)/2; end;
       for i=1:n(4)+1
           V_4(:,i) = V_4(:,i) * sqrt(4*i-2)/2;
           Vx(:,i) = Vx(:,i) * sqrt(4*i-2)/2;
90         Vl(:,i) = Vl(:,i) * sqrt(4*i-2)/2;
           Vr(:,i) = Vr(:,i) * sqrt(4*i-2)/2;
       end;
       V_1_inv = inv(V_1); V_2_inv = inv(V_2); V_3_inv = inv(V_3); V_4_inv = inv(V_4);
       Vcond = [ cond(V_1) , cond(V_2) , cond(V_3) , cond(V_4) ];
95
       % shift matrix
```

```
        T_left = V_4_inv * Vl;
        T_right = V_4_inv * Vr;

100     % compute the integral
        % w := [ seq( int(T(k,x),x=-1..1) , k=0..n ) ];
        w = [ sqrt(2) , zeros(1,n(4)) ] / 2; % legendre

        % set-up the downdate matrix
105     k = [0:n(4)]';
        U = diag(sqrt((k+1).^2 ./ (2*k+1) ./ (2*k+3))) + diag(sqrt(k(3:end).^2 ./ (4*k(3:end).^2-1)),2);

        end; % if exist('n')

110     % create the original datatype
        ivals = struct( ...
            'a', [], 'b',[], ...
            'c', [], 'c_old', [], ...
            'fx', [], ...
115         'int', [], ...
            'err', [], ...
            'tol', [], ...
            'depth', [], ...
            'rdepth', [], ...
120         'ndiv', [] );

        % compute the first interval
        points = (a+b)/2 + (b-a) * xi_4 / 2;
        fx = f(points); nans = [];
125     for i=1:length(fx), if ~isfinite(fx(i)), nans = [ i , nans ]; fx(i) = 0.0; end; end;
        ivals(1).c = zeros(n(4)+1,4);
        ivals(1).c(1:n(4)+1,4) = V_4_inv * fx;
        ivals(1).c(1:n(3)+1,3) = V_3_inv * fx([1:2:n(4)+1]);
        for i=nans, fx(i) = NaN; end;
130     ivals(1).fx = fx;
        ivals(1).c_old = zeros(size(fx));
        ivals(1).a = a; ivals(1).b = b;
        ivals(1).int = (b-a) * w * ivals(1).c(:,4);
        c_diff = norm(ivals(1).c(:,4) - ivals(1).c(:,3));
135     ivals(1).err = (b-a) * c_diff;
        if c_diff / norm(ivals(1).c(:,4)) > 0.1
            ivals(1).err = max( ivals(1).err , (b-a) * norm(ivals(1).c(:,4)) );
        end;
        ivals(1).tol = tol;
140     ivals(1).depth = 4;
        ivals(1).ndiv = 0;
        ivals(1).rdepth = 1;

        % init some globals
145     int = ivals(1).int; err = ivals(1).err; nr_ivals = 1;
        int_final = 0; err_final = 0; err_excess = 0;
        i_max = 1; nr_points = n(4)+1;
        ndiv_max = 20;

150     % do we even need to go this way?
        if err < int * tol, return; end;

        % main loop
        while true
155
            % get some global data
            a = ivals(i_max).a; b = ivals(i_max).b;
            depth = ivals(i_max).depth;
            split = 0;
160
            % depth of the old interval
            if depth == 1
                points = (a+b)/2 + (b-a)*xi_2/2;
                fx(1:2:n(2)+1) = ivals(i_max).fx;
165             fx(2:2:n(2)) = f(points(2:2:n(2)));
                fx = fx(1:n(2)+1);
                nans = [];
                for i=1:length(fx), if ~isfinite(fx(i)), fx(i) = 0.0; nans = [ i , nans ]; end; end;
                c_new = V_2_inv * fx;
170             if length(nans) > 0
                    b_new = b_2_def; n_new = n(2);
```

```
         for i=nans
             b_new(1:end-1) = (U(1:n(2)+1,1:n(2)+1) - diag(ones(n(2),1)*xi_2(i),1)) \ b_new(2:end);
             b_new(end) = 0; c_new = c_new - c_new(n_new+1) / b_new(n_new+1) * b_new(1:end-1);
175          n_new = n_new - 1; fx(i) = NaN;
         end;
       end;
       ivals(i_max).fx = fx;
       nc = norm(c_new);
180    nr_points = nr_points + n(2)-n(1);
       ivals(i_max).c(1:n(2)+1,2) = c_new;
       c_diff = norm(ivals(i_max).c(:,1) - ivals(i_max).c(:,2));
       ivals(i_max).err = (b-a) * c_diff;
       int_old = ivals(i_max).int; ivals(i_max).int = (b-a) * w(1) * c_new(1);
185    if nc > 0 && c_diff / nc > 0.1
           split = 1;
       else
           ivals(i_max).depth = 2;
       end;
190    elseif depth == 2
       points = (a+b)/2 + (b-a)*xi_3/2;
       fx(1:2:n(3)+1) = ivals(i_max).fx;
       fx(2:2:n(3)) = f(points(2:2:n(3)));
       fx = fx(1:n(3)+1);
195    nans = [];
       for i=1:length(fx), if ~isfinite(fx(i)), fx(i) = 0.0; nans = [ i , nans ]; end; end;
       c_new = V_3_inv * fx;
       if length(nans) > 0
           b_new = b_3_def; n_new = n(3);
200        for i=nans
               b_new(1:end-1) = (U(1:n(3)+1,1:n(3)+1) - diag(ones(n(3),1)*xi_3(i),1)) \ b_new(2:end);
               b_new(end) = 0; c_new = c_new - c_new(n_new+1) / b_new(n_new+1) * b_new(1:end-1);
               n_new = n_new - 1; fx(i) = NaN;
           end;
205    end;
       ivals(i_max).fx = fx;
       nc = norm(c_new);
       nr_points = nr_points + n(3)-n(2);
       ivals(i_max).c(1:n(3)+1,3) = c_new;
210    c_diff = norm(ivals(i_max).c(:,2) - ivals(i_max).c(:,3));
       ivals(i_max).err = (b-a) * c_diff;
       int_old = ivals(i_max).int; ivals(i_max).int = (b-a) * w(1) * c_new(1);
       if nc > 0 && c_diff / nc > 0.1
           split = 1;
215    else
           ivals(i_max).depth = 3;
       end;
       elseif depth == 3
       points = (a+b)/2 + (b-a)*xi_4/2;
220    fx(1:2:n(4)+1) = ivals(i_max).fx;
       fx(2:2:n(4)) = f(points(2:2:n(4)));
       fx = fx(1:n(4)+1);
       nans = [];
       for i=1:length(fx), if ~isfinite(fx(i)), fx(i) = 0.0; nans = [ i , nans ]; end; end;
225    c_new = V_4_inv * fx;
       if length(nans) > 0
           b_new = b_4_def; n_new = n(4);
           for i=nans
               b_new(1:end-1) = (U(1:n(4)+1,1:n(4)+1) - diag(ones(n(4),1)*xi_4(i),1)) \ b_new(2:end);
230            b_new(end) = 0; c_new = c_new - c_new(n_new+1) / b_new(n_new+1) * b_new(1:end-1);
               n_new = n_new - 1; fx(i) = NaN;
           end;
       end;
       ivals(i_max).fx = fx;
235    nc = norm(c_new);
       nr_points = nr_points + n(4)-n(3);
       ivals(i_max).c(:,4) = c_new;
       c_diff = norm(ivals(i_max).c(:,3) - ivals(i_max).c(:,4));
       ivals(i_max).err = (b-a) * c_diff;
240    int_old = ivals(i_max).int; ivals(i_max).int = (b-a) * w(1) * c_new(1);
       if nc > 0 && c_diff / nc > 0.1
           split = 1;
       else
           ivals(i_max).depth = 4;
245    end;
       else
```

```
               split = 1;
             end;
250    % can we safely ignore this interval?
       if points(2) <= points(1) || points(end) <= points(end-1) || ...
          ivals(i_max).err < abs(ivals(i_max).int) * eps * Vcond(ivals(i_max).depth)

             err_final = err_final + ivals(i_max).err;
255          int_final = int_final + ivals(i_max).int;
             ivals(i_max) = ivals(nr_ivals);
             nr_ivals = nr_ivals - 1;
       elseif split
             m = (a + b) / 2;
260          nr_ivals = nr_ivals + 1;
             ivals(nr_ivals).a = a; ivals(nr_ivals).b = m;
             ivals(nr_ivals).tol = ivals(i_max).tol / sqrt(2);
             ivals(nr_ivals).depth = 1; ivals(nr_ivals).rdepth = ivals(i_max).rdepth + 1;
             ivals(nr_ivals).c = zeros(n(4)+1,4);
265          fx = [ ivals(i_max).fx(1) ; f((a+m)/2+(m-a)*xi_1(2:end-1)/2) ; ivals(i_max).fx((end+1)/2) ];
             nr_points = nr_points + n(1)-1; nans = [];
             for i=1:length(fx), if ~isfinite(fx(i)), fx(i) = 0.0; nans = [ i , nans ]; end; end;
             c_new = V_1_inv * fx;
             if length(nans) > 0
270              b_new = b_1_def; n_new = n(1);
                 for i=nans
                     b_new(1:end-1) = (U(1:n(1)+1,1:n(1)+1) - diag(ones(n(1),1)*xi_1(i),1)) \ b_new(2:end);
                     b_new(end) = 0; c_new = c_new - c_new(n_new+1) / b_new(n_new+1) * b_new(1:end-1);
                     n_new = n_new - 1; fx(i) = NaN;
275              end;
             end;
             ivals(nr_ivals).fx = fx;
             ivals(nr_ivals).c(1:n(1)+1,1) = c_new; nc = norm(c_new);
             ivals(nr_ivals).c_old = T_left * ivals(i_max).c(:,depth);
280          c_diff = norm(ivals(nr_ivals).c(:,1) - ivals(nr_ivals).c_old);
             ivals(nr_ivals).err = (m - a) * c_diff;
             ivals(nr_ivals).int = (m - a) * ivals(nr_ivals).c(1,1) * w(1);
             ivals(nr_ivals).ndiv = ivals(i_max).ndiv + ...
                 (abs(ivals(i_max).c(1,1)) > 0 && ivals(nr_ivals).c(1,1) / ivals(i_max).c(1,1) > 2);
285          if ivals(nr_ivals).ndiv > ndiv_max && 2*ivals(nr_ivals).ndiv > ivals(nr_ivals).rdepth
                 warning(sprintf('possibly␣divergent␣integral␣in␣the␣interval␣[%e,%e]!␣(h=%e)',a,m,m-a));
                 int = sign(int) * Inf;
                 return;
             end;
290          nr_ivals = nr_ivals + 1;
             ivals(nr_ivals).a = m; ivals(nr_ivals).b = b;
             ivals(nr_ivals).tol = ivals(i_max).tol / sqrt(2);
             ivals(nr_ivals).depth = 1; ivals(nr_ivals).rdepth = ivals(i_max).rdepth + 1;
             ivals(nr_ivals).c = zeros(n(4)+1,4);
295          fx = [ ivals(i_max).fx((end+1)/2) ; f((m+b)/2+(b-m)*xi_1(2:end-1)/2) ; ivals(i_max).fx(end) ];
             nr_points = nr_points + n(1)-1; nans = [];
             for i=1:length(fx), if ~isfinite(fx(i)), fx(i) = 0.0; nans = [ i , nans ]; end; end;
             c_new = V_1_inv * fx;
             if length(nans) > 0
300              b_new = b_1_def; n_new = n(1);
                 for i=nans
                     b_new(1:end-1) = (U(1:n(1)+1,1:n(1)+1) - diag(ones(n(1),1)*xi_1(i),1)) \ b_new(2:end);
                     b_new(end) = 0; c_new = c_new - c_new(n_new+1) / b_new(n_new+1) * b_new(1:end-1);
                     n_new = n_new - 1; fx(i) = NaN;
305              end;
             end;
             ivals(nr_ivals).fx = fx;
             ivals(nr_ivals).c(1:n(1)+1,1) = c_new; nc = norm(c_new);
             ivals(nr_ivals).c_old = T_right * ivals(i_max).c(:,depth);
310          c_diff = norm(ivals(nr_ivals).c(:,1) - ivals(nr_ivals).c_old);
             ivals(nr_ivals).err = (b - m) * c_diff;
             ivals(nr_ivals).int = (b - m) * ivals(nr_ivals).c(1,1) * w(1);
             ivals(nr_ivals).ndiv = ivals(i_max).ndiv + ...
                 (abs(ivals(i_max).c(1,1)) > 0 && ivals(nr_ivals).c(1,1) / ivals(i_max).c(1,1) > 2);
315          if ivals(nr_ivals).ndiv > ndiv_max && 2*ivals(nr_ivals).ndiv > ivals(nr_ivals).rdepth
                 warning(sprintf('possibly␣divergent␣integral␣in␣the␣interval␣[%e,%e]!␣(h=%e)',m,b,b-m));
                 int = sign(int) * Inf;
                 return;
             end;
320          ivals(i_max) = ivals(nr_ivals);
             nr_ivals = nr_ivals - 1;
```

210   ALGORITHMS IN MATLAB


```
       end;

       % compute the running err and new max
325    i_max = 1; i_min = 1; err = err_final; int = int_final;
       for i=1:nr_ivals
          if ivals(i).err > ivals(i_max).err, i_max = i;
          elseif ivals(i).err < ivals(i_min).err, i_min = i; end;
          err = err + ivals(i).err;
330       int = int + ivals(i).int;
       end;

       % nuke smallest element if stack is larger than 500
       if nr_ivals > 200
335       err_final = err_final + ivals(i_min).err;
          int_final = int_final + ivals(i_min).int;
          ivals(i_min) = ivals(nr_ivals);
          if i_max == nr_ivals, i_max = i_min; end;
          nr_ivals = nr_ivals - 1;
340    end;

       % get up and leave?
       if err < abs(int) * tol || ...
          (err_final > abs(int) * tol && err - err_final < abs(int) * tol) || ...
345       nr_ivals == 0
          break;
       end;

    end; % main loop
350
    % clean-up and tabulate
    err = err + err_excess;

end
```

B.3   An Adaptive Quadrature Algorithm using the Refined Error Estimate

```
   function [ int , err , nr_points ] = int_gonnet2008b ( f , a , b , tol )
   %INT_GONNET2008B  evaluates an integral using adaptive quadrature. The
   %    algorithm is based on a 10-point Clenshaw-Curtis rule and the error
   %    is extrapolated from the L2-norm of the difference between two
 5 %    successive interpolations of the integrand over the nodes of the
   %    respective quadrature rules.
   %
   %    INT = INT_GONNET2008B ( F , A , B , TOL ) approximates the integral
   %    of F in the interval [A,B] up to the relative tolerance TOL. The
10 %    integrand F should accept a vector argument and return a vector
   %    result containing the integrand evaluated at each element of the
   %    argument.
   %
   %    [INT,ERR,NR_POINTS] = INT_GONNET2008B ( F , A , B , TOL ) returns
15 %    ERR, an estimate of the absolute integration error as well as
   %    NR_POINTS, the number of function values for which the integrand
   %    was evaluated. The value of ERR may be larger than the requested
   %    tolerance, indicating that the integration may have failed.
   %
20 %    INT_GONNET2008B halts with a warning if the integral is or appears
   %    to be divergent.
   %
   %    Reference: "Adaptive Quadrature Re-Revisited", P. Gonnet, submitted
   %        to ACM Transactions on Mathematical Software, 2008.
25
       % declare persistent variables
       persistent n xi b_def b_def_left b_def_right V Vx xx Vinv Vcond Tleft Tright w

       % have the persistent variables been declared already?
30     if ~exist('w') || isempty(w)

           % the nodes and function values
```

```
      n = 10;
      xi = -cos([0:n]/n*pi)';
35    b_def = [0., .10620845052443988729141015545646e-4, 0., ...
      .22713052318703509770686120429e-4, 0., .58188804291554955852478875454e-4, ...
      0., .33466729598031323672079288921e-3, 0., -.13015591094231324681626313827e-2, 0., ...
      .85610271778209474285779687247e-3]';
      b_def_left = 2^(n+1) * [.276213586400995126719079828971e-3, ...
40    .499657655492227449762439076973e-3, .523434680236984721004073014910e-3, ...
      .257874261861280205575039917099e-3, -.748342903416082700961183312797e-3, ...
      -.917669983469695220709693581407e-3, .549500980250147994974675357971e-3, ...
      .300060029460695461688333929627e-3, -.276678492776836119682157766061e-3, ...
45    .761043473405371645812866738709e-4, -.918691057168920637420186132198e-5, ...
      .418018905167038448661033629139e-6]';
      b_def_right = 2^(n+1) * [-.276213586400995126719079828971e-3, ...
      .499657655492227449762439076973e-3, -.523434680236984721004073014910e-3, ...
      .257874261861280205575039917099e-3, -.748342903416082700961183312797e-3, ...
50    -.917669983469695220709693581407e-3, -.549500980250147994974675357971e-3, ...
      .300060029460695461688333929627e-3, .276678492776836119682157766061e-3, ...
      .761043473405371645812866738709e-4, .918691057168920637420186132198e-5, ...
      .418018905167038448661033629139e-6]';

      % compute the coefficients
55    V = [ ones(size(xi)) xi ];
      xil = (xi - 1) / 2; xir = (xi + 1) / 2;
      Vl = [ ones(size(xil)) xil ]; Vr = [ ones(size(xir)) xir ];
      xx = linspace(-1,1,200)'; Vx = [ ones(size(xx)) xx ];
      for i=3:n+2
60      V = [ V , ((2*i-3) / (i-1) * xi .* V(:,i-1) - (i-2) / (i-1) * V(:,i-2)) ];
        Vx = [ Vx , ((2*i-3) / (i-1) * xx .* Vx(:,i-1) - (i-2) / (i-1) * Vx(:,i-2)) ];
        Vl = [ Vl , ((2*i-3) / (i-1) * xil .* Vl(:,i-1) - (i-2) / (i-1) * Vl(:,i-2)) ];
        Vr = [ Vr , ((2*i-3) / (i-1) * xir .* Vr(:,i-1) - (i-2) / (i-1) * Vr(:,i-2)) ];
      end;
65    for i=1:n+2
        V(:,i) = V(:,i) * sqrt(4*i-2)/2;
        Vx(:,i) = Vx(:,i) * sqrt(4*i-2)/2;
        Vl(:,i) = Vl(:,i) * sqrt(4*i-2)/2;
        Vr(:,i) = Vr(:,i) * sqrt(4*i-2)/2;
70    end;
      Vinv = inv(V(:,1:n+1));
      Vcond = cond(V);

      % shift matrix
75    Tleft = V(:,1:n+1) \ Vl(:,1:n+1);
      Tright = V(:,1:n+1) \ Vr(:,1:n+1);

      % compute the integral
      w = [ sqrt(2) , zeros(1,n) ] / 2; % legendre
80
      end; % if exist('w')

      % create the original datatype
      ivals = struct( ...
85    'a', [], 'b',[], ...
      'fx', [], ...
      'c', [], 'c_old', [], ...
      'b_new', [], 'b_old', [], ...
      'fn', [], 'fn_old', [], ...
90    'int', [], ...
      'err', [], ...
      'tol', [], ...
      'depth', [], ...
      'ndiv' , [] );
95    ivals_new = ivals;

      % compute the first interval
      points = a + (b-a) * (xi + 1) / 2;
      fx = f(points); nans = [];
100   for i=1:length(fx), if ~isfinite(fx(i)), nans = [i,nans]; fx(i) = 0.0; end; end;
      ivals(1).c = Vinv * fx;
      for i=nans, fx(i) = NaN; end;
      ivals(1).fx = fx;
      ivals(1).c_old = zeros(size(fx));
105   ivals(1).a = a; ivals(1).b = b;
      ivals(1).int = (b-a) * w * ivals(1).c;
      ivals(1).b_new = b_def;
```

```
      ivals(1).fn = 0;
      ivals(1).err = abs((b-a) * norm(ivals(1).c));
110   ivals(1).tol = tol;
      ivals(1).depth = 1;
      ivals(1).ndiv = 0;

      % init some globals
115   int = ivals(1).int; err = ivals(1).err; nr_ivals = 1;
      int_final = 0; err_final = 0; err_excess = 0;
      i_max = 1; nr_points = n+1;
      ndiv_max = 20; theta_1 = 1.1;

120   % main loop
      while true

          % compute the left interval
          a_left = ivals(i_max).a; b_left = (ivals(i_max).a + ivals(i_max).b)/2;
125       h_left = b_left - a_left;
          points_left = a_left + h_left * (xi + 1) / 2;
          nr_points = nr_points + n-2;
          fx_left = [ ivals(i_max).fx(1) ; f(points_left(2:end-1)) ; ivals(i_max).fx(n/2+1) ];
          nans = [];
130       for i=1:length(fx_left), if ~isfinite(fx_left(i)), nans = [ nans , i ]; fx_left(i) = 0.0; end; end;
          c_left = Vinv * fx_left;
          if length(nans) > 0
              k = [0:length(xi)-1]';
              U = diag(sqrt((k+1).^2 ./ (2*k+1) ./ (2*k+3))) + diag(sqrt(k(3:end).^2 ./ (4*k(3:end).^2-1)),2);
135           b_new_left = b_def; n_left = n;
              for i=nans
                  b_new_left(1:end-1) = (U - diag(ones(length(xi)-1,1)*xi(i),1)) \ b_new_left(2:end);
                  b_new_left(end) = 0;
                  c_left = c_left - c_left(n_left+1) / b_new_left(n_left+1) * b_new_left(1:end-1);
140               n_left = n_left - 1; fx_left(i) = NaN;
              end;
          else
              b_new_left = b_def;
          end;
145       nc_left = norm(c_left);
          c_old_left = Tleft * ivals(i_max).c;
          c_diff = norm(c_left - c_old_left);
          if ivals(i_max).b_new(end) == 0, b_old_left =  2^(n+1) * [ Tleft * ivals(i_max).b_new(1:end-1) ; 0 ];
          else b_old_left = b_def_left; end;
150       fn_left = c_diff / norm(b_old_left - b_new_left);
          int_left = h_left * w * c_left;
          if abs(ivals(i_max).int) > 0 && int_left / ivals(i_max).int > 1
              ndiv_left = ivals(i_max).ndiv + 1;
          else
155           ndiv_left = ivals(i_max).ndiv;
          end;
          if ndiv_left > ndiv_max && 2*ndiv_right > ivals(i_max).depth
              warning(sprintf('possibly_divergent_integral_in_the_interval_[%e,%e]!',a_left,b_left));
              int = sign(int) * Inf;
160           return;
          end;
          absVcmf = abs(V(:,1:n+1) * c_old_left - fx_left);
          absVb = abs( V * b_old_left );
          if max(absVcmf(2:end-1) - theta_1 * fn_left * absVb(2:end-1)) > eps*Vcond*nc_left
165           err_left = h_left * c_diff;
          else
              err_left = abs( h_left * fn_left * norm(b_new_left) );
          end;

170       % store the left interval
          if points_left(end) <= points_left(end-1) || ...
              err_left < abs(int_left) * Vcond * eps
              % disp(sprintf('excess: %e (eps=%e,h=%e)',err_left,nc_left * Vcond * eps,b_left - a_left));
              err_excess = err_excess + err_left;
175           int_final = int_final + int_left;
          else % if err_left > abs(int) * ivals(i_max).tol / sqrt(2)
              nr_ivals = nr_ivals + 1;
              ivals(nr_ivals).c = c_left;
              ivals(nr_ivals).fx = fx_left;
180           ivals(nr_ivals).c_old = c_old_left;
              ivals(nr_ivals).a = a_left; ivals(nr_ivals).b = b_left;
              ivals(nr_ivals).b_new = b_new_left;
```

```
            ivals(nr_ivals).b_old = b_old_left;
            ivals(nr_ivals).fn = fn_left;
185         ivals(nr_ivals).fn_old = ivals(i_max).fn;
            ivals(nr_ivals).int = int_left;
            ivals(nr_ivals).err = err_left;
            ivals(nr_ivals).tol = ivals(i_max).tol / sqrt(2);
            ivals(nr_ivals).depth = ivals(i_max).depth + 1;
190         ivals(nr_ivals).ndiv = ndiv_left;
          end;

          % compute the right interval
          a_right = (ivals(i_max).a + ivals(i_max).b)/2; b_right = ivals(i_max).b;
195       h_right = b_right - a_right;
          points_right = a_right + h_right * (xi + 1) / 2;
          nr_points = nr_points + n-2;
          fx_right = [ ivals(i_max).fx(n/2+1) ; f(points_right(2:end-1)) ; ivals(i_max).fx(end) ];
          nans = [];
200       for i=1:length(fx_right), if ~isfinite(fx_right(i)), nans = [ nans , i ]; fx_right(i) = 0.0; end; end;
          c_right = Vinv * fx_right;
          if length(nans) > 0
              k = [0:length(xi)-1]';
              U = diag(sqrt((k+1).^2 ./ (2*k+1) ./ (2*k+3))) + diag(sqrt(k(3:end).^2 ./ (4*k(3:end).^2-1)),2);
205           b_new_right = b_def; n_right = n;
              for i=nans
                  b_new_right(1:end-1) = (U - diag(ones(length(xi)-1,1)*xi(i),1)) \ b_new_right(2:end);
                  b_new_right(end) = 0;
                  c_right = c_right - c_right(n_right+1) / b_new_right(n_right+1) * b_new_right(1:end-1);
210               n_right = n_right - 1; fx_right(i) = NaN;
              end;
          else
              b_new_right = b_def;
          end;
215       nc_right = norm(c_right);
          c_old_right = Tright * ivals(i_max).c;
          c_diff = norm(c_right - c_old_right);
          if ivals(i_max).b_new(end) == 0, b_old_right = 2^(n+1) * [ Tright * ivals(i_max).b_new(1:end-1) ; 0 ];
          else b_old_right = b_def_right; end;
220       fn_right = c_diff / norm(b_old_right - b_new_right);
          int_right = h_right * w * c_right;
          if abs(ivals(i_max).int) > 0 && int_right / ivals(i_max).int > 1
              ndiv_right = ivals(i_max).ndiv + 1;
          else
225           ndiv_right = ivals(i_max).ndiv;
          end;
          if ndiv_right > ndiv_max && 2*ndiv_right > ivals(i_max).depth
              warning(sprintf('possibly divergent integral in the interval [%e,%e]!',a_right,b_right));
              int = sign(int) * Inf;
230           return;
          end;
          absVcmf = abs(V(:,1:n+1) * c_old_right - fx_right);
          absVb = abs( V * b_old_right );
          if max(absVcmf(2:end-1) - theta_1 * fn_right * absVb(2:end-1)) > eps*Vcond*nc_left
235           err_right = h_right * c_diff;
          else
              err_right = abs( h_right * fn_right * norm(b_new_right) );
          end;

240       % store the right interval
          if points_right(end) <= points_right(end-1) || ...
              err_right < abs(int_right) * Vcond * eps
              % disp(sprintf('excess: %e (eps=%e,h=%e)',err_right,nc_right * Vcond * eps,b_right - a_right));
              err_excess = err_excess + err_right;
245           int_final = int_final + int_right;
          else % if err_right > abs(int) * ivals(i_max).tol / sqrt(2)
              nr_ivals = nr_ivals + 1;
              ivals(nr_ivals).c = c_right;
              ivals(nr_ivals).fx = fx_right;
250           ivals(nr_ivals).c_old = c_old_right;
              ivals(nr_ivals).a = a_right; ivals(nr_ivals).b = b_right;
              ivals(nr_ivals).b_new = b_new_right;
              ivals(nr_ivals).b_old = b_old_right;
              ivals(nr_ivals).fn = fn_right;
255           ivals(nr_ivals).fn_old = ivals(i_max).fn;
              ivals(nr_ivals).int = int_right;
              ivals(nr_ivals).err = err_right;
```

```
            ivals(nr_ivals).tol = ivals(i_max).tol / sqrt(2);
            ivals(nr_ivals).depth = ivals(i_max).depth + 1;
260         ivals(nr_ivals).ndiv = ndiv_right;
        end;

        % copy last interval into i_max
        ivals(i_max) = ivals(nr_ivals);
265     nr_ivals = nr_ivals - 1;

        % compute the running err and new max
        i_max = 1; i_min = 1; err = err_final; int = int_final;
        for i=1:nr_ivals
270         if ivals(i).err > ivals(i_max).err, i_max = i;
            elseif ivals(i).err < ivals(i_min).err, i_min = i; end;
            err = err + ivals(i).err;
            int = int + ivals(i).int;
        end;
275
        % nuke smallest element if stack is larger than 500
        if nr_ivals > 200
            err_final = err_final + ivals(i_min).err;
            int_final = int_final + ivals(i_min).int;
280         ivals(i_min) = ivals(nr_ivals);
            if i_max == nr_ivals, i_max = i_min; end;
            nr_ivals = nr_ivals - 1;
        end;

285     % are we done yet?
        if err < abs(int) * tol || ...
           (err_final > abs(int) * tol && err - err_final < abs(int) * tol) || ...
           nr_ivals == 0
            break;
290     end;

    end; % main loop

        % clean-up and tabulate
295     err = err + err_excess;

end
```

Bibliography

[1] A. Aitken. On Bernoulli's numerical solution of algebraic equations. *Proceedings of the Royal Society of Edinburgh*, 46:289–305, 1926.

[2] Z. Balles and L. N. Trefethen. An extensions of MATLAB to continuous functions and operators. *SIAM Journal of Scientific Computing*, 25(5):1743–1770, 2004.

[3] F. L. Bauer, H. Rutishauser, and E. Stiefel. New aspects in numerical quadrature. In N. C. Metropolis, A. H. Taub, J. Todd, and C. B. Tompkins, editors, *Experimental Arithmetic, High Speed Computing and Mathematics*, pages 199–217, Providence, RI, 1963. American Mathematical Society.

[4] J. Berntsen and T. O. Espelid. On the use of Gauss quadrature in adaptive automatic integration schemes. *BIT Numerical Mathematics*, 24:239–242, 1984.

[5] J. Berntsen and T. O. Espelid. Error estimation in automatic quadrature routines. *ACM Transactions on Mathematical Software*, 17(2):233–252, 1991.

[6] J. P. Berrut and L. N. Trefethen. Barycentric Lagrange interpolation. *SIAM Review*, 46(3):501–517, 2004.

[7] Å. Björck and V. Pereyra. Solution of Vandermonde systems of equations. *Mathematics of Computation*, 24(112):893–903, 1970.

[8] J. Casaletto, M. Pickett, and J. Rice. A comparison of some numerical integration programs. *SIGNUM Newsletter*, 4(3):30–40, 1969.

[9] B. Char, K. Geddes, and G. Gonnet. The Maple symbolic computation system. *SIGSAM Bull.*, 17(3-4):31–42, 1983.

[10] C. W. Clenshaw and A. R. Curtis. A method for numerical integration on an automatic computer. *Numerische Mathematik*, 2:197–205, 1960.

[11] R. Cools and A. Haegemans. CUBPACK: Progress report. In T. O. Espelid and A. Genz, editors, *Numerical Integration*, pages 305–315. Kluwer Academic Publishers, Netherlands, 1992.

[12] P. J. Davis and P. Rabinowitz. *Numerical Integration.* Blaisdell Publishing Company, Waltham, Massachusetts, 1967.

[13] P. J. Davis and P. Rabinowitz. *Numerical Integration, 2nd Edition.* Blaisdell Publishing Company, Waltham, Massachusetts, 1984.

[14] C. de Boor. CADRE: An algorithm for numerical quadrature. In J. R. Rice, editor, *Mathematical Software*, pages 201–209. Academic Press, New York and London, 1971.

[15] E. de Doncker. An adaptive extrapolation algorithm for automatic integration. *SIGNUM Newsletter*, 13(2):12–18, 1978.

[16] J. W. Eaton. *GNU Octave Manual.* Network Theory Limited, 2002.

[17] H. Engels. *Numerical quadrature and cubature.* Academic Press, London, New York, 1980.

[18] T. O. Espelid. Integration rules, null rules and error estimation. Technical Report 22, Department of Informatics, University of Bergen, Norway, December 1988.

[19] T. O. Espelid. DQUAINT: An algorithm for adaptive quadrature over a collection of finite intervals. In T. O. Espelid and A. Genz, editors, *Numerical Integration*, pages 341–342. Kluwer Academic Publishers, Netherlands, 1992.

[20] T. O. Espelid. Doubly adaptive quadrature routines based on Newton-Cotes rules. Technical Report 229, Department of Informatics, University of Bergen, Norway, May 2002.

[21] T. O. Espelid. Doubly adaptive quadrature routines based on Newton-Cotes rules. *BIT Numerical Mathematics*, 43:319–337, 2003.

[22] T. O. Espelid. Extended doubly adaptive quadrature routines. Technical Report 266, Department of Informatics, University of Bergen, Norway, February 2004.

[23] T. O. Espelid. A test of QUADPACK and four doubly adaptive quadrature routines. Technical Report 281, Department of Informatics, University of Bergen, Norway, October 2004.

[24] T. O. Espelid. Algorithm 868: Globally doubly adaptive quadrature – reliable Matlab codes. *ACM Transactions on Mathematical Software*, 33(3):Article 21, 2007.

[25] T. O. Espelid and T. Sørevik. A discussion of a new error estimate for adaptive quadrature. *BIT Numerical Mathematics*, 29:293–294, 1989.

[26] P. Favati, G. Lotti, and F. Romani. Interpolatory integration formulas for optimal composition. *ACM Transactions on Mathematical Software*, 17(2):207–217, 1991.

[27] P. Favati, G. Lotti, and F. Romani. Local error estimates in quadrature. *BIT Numerical Mathematics*, 31:102–111, 1991.

[28] L. Fejér. Mechanische Quadraturen mit positiven Cotesschen Zahlen. *Mathematische Zeitschrift*, 37:287–309, 1933.

[29] G. E. Forsythe, M. A. Malcolm, and C. B. Moler. *Computer Methods for Mathematical Computations*. Prentice-Hall, Inc., Englewood Cliffs, N.J. 07632, 1977.

[30] L. J. Gallaher. Algorithm 303: An adaptive quadrature procedure with random panel sizes. *Communications of the ACM*, 10(6):373–374, 1967.

[31] W. Gander. Change of basis in polynomial interpolation. *Numerical Linear Algebra with Applications*, 12(8):769–778, 2005.

[32] W. Gander. Generating numerical algorithms using a computer algebra system. *BIT Numerical Mathematics*, 46:491–504, 2006.

[33] W. Gander and W. Gautschi. Adaptive quadrature — revisited. Technical Report 306, Department of Computer Science, ETH Zurich, Switzerland, 1998.

[34] W. Gander and W. Gautschi. Adaptive quadrature — revisited. *BIT*, 40(1):84–101, 2001.

[35] S. Garribba, L. Quartapelle, and G. Reina. SNIFF: Efficient self-tuning algorithm for numerical integration. *Computing*, 20:363–375, 1978.

[36] W. Gautschi. Norm estimates for inverses of Vandermonde matrices. *Numerische Mathematik*, 23:337–347, 1975.

[37] W. Gautschi. The condition of Vandermonde-like matrices involving orthogonal polynomials. *Linear Algebra and its Applications*, 52–53:293–300, 1983.

[38] W. Gautschi. *Numerical Analysis, An Introduction*. Birkhäuser Verlag, Boston, Basel and Stuttgart, 1997.

[39] W. Gautschi. *Orthogonal Polynomials: Computation and Approximation*. Oxford University Press, Great Clarendon Street, Oxford OX2 6DP, 2004.

[40] W. M. Gentleman. Implementing Clenshaw-Curtis quadrature I and II. *Communications of the ACM*, 15(5):337–346, 1972.

[41] G. H. Golub and C. F. Van Loan. *Matrix Computations*. John Hopkins University Press, Baltimore, Maryland, third edition, 1996.

[42] T. Hasegawa, S. Hibino, Y. Hosoda, and I. Ninomiya. An extended doubly-adaptive quadrature method based on the combination of the Ninomiya and the FLR schemes. *Numerical Algorithms*, 45(1–4):101–112, 2007.

[43] P. Henrici. *Elements of numerical analysis*. John Wiley & Sons Inc., New York, 1964.

[44] P. Henrici. *Essentials of Numerical Analysis, with Pocket Calculator Demonstrations*. John Wiley & Sons, Inc., New York, 1982.

[45] S. Henriksson. Contribution no. 2: Simpson numerical integration with variable length of step. *BIT Numerical Mathematics*, 1:290, 1961.

[46] N. J. Higham. Fast solution of Vandermonde-like systems involving orthogonal polynomials. *IMA Journal of Numerical Analysis*, 8:473–486, 1988.

[47] N. J. Higham. Stability analysis of algorithms for solving confluent Vandermonde-like systems. *SIAM Journal on Matrix Analysis and Applications*, 11(1):23–41, 1990.

[48] K. E. Hillstrom. Comparison of several adaptive Newton-Cotes quadrature routines in evaluating definite integrals with peaked integrands. *Communications of the ACM*, 13(6):362–365, 1970.

[49] D. K. Kahaner. 5.15 Comparison of numerical quadrature formulas. In J. R. Rice, editor, *Mathematical Software*, pages 229–259. Academic Press, New York and London, 1971.

[50] D. K. Kahaner. Numberical quadrature by the ε-algorithm. *Mathematics of Computation*, 26(119):689–693, 1972.

[51] D. K. Kahaner and J. Stoer. Extrapolated adaptive quadrature. *SIAM Journal on Scientific and Statistical Computing*, 4(1):31–44, 1983.

[52] F. T. Krogh and W. Van Snyder. Algorithm 699: A new representation of Patterson's quadrature formulae. *ACM Transactions on Mathematical Software*, 17(4):457–461, 1991.

[53] A. R. Krommer and C. W. Überhuber. *Computational Integration*. SIAM, Philadelphia, 1998.

[54] A. S. Kronrod. *Nodes and Weights of Quadrature Formulas — Authorized Translation from the Russian*. Consultants Bureau, New York, 1965.

[55] G. F. Kuncir. Algorithm 103: Simpson's rule integrator. *Communications of the ACM*, 5(6):347, 1962.

[56] D. P. Laurie. Sharper error estimates in adaptive quadrature. *BIT Numerical Mathematics*, 23:258–261, 1983.

[57] D. P. Laurie. Practical error estimation in numerical integration. *Journal of Computational and Applied Mathematics*, 12–13:425–431, 1985.

[58] D. P. Laurie. Stratified sequences of nested quadrature formulas. *Quaestiones Mathematicae*, 15:364–384, 1992.

[59] D. P. Laurie. Null rules and orthogonal expansions. In *Approximation and computation (West Lafayette, IN, 1993)*, volume 119 of *Internat. Ser. Numer. Math.*, pages 359–370. Birkhäuser Boston, Boston, MA, 1994.

[60] J. N. Lyness. Symmetric integration rules for hypercubes III: Construction of integration rules using null rules. *Mathematics of Computation*, 19:625–637, 1965.

[61] J. N. Lyness. Notes on the adaptive Simpson quadrature routine. *Journal of the ACM*, 16(3):483–495, 1969.

[62] J. N. Lyness. Algorithm 379: SQUANK (Simpson Quadrature Used Adaptively – Noise Killed). *Communications of the ACM*, 13(4):260–262, 1970.

[63] J. N. Lyness and J. J. Kaganove. Comments on the nature of automatic quadrature routines. *ACM Transactions on Mathematical Software*, 2(1):170–177, 1976.

[64] J. N. Lyness and J. J. Kaganove. A technique for comparing automatic quadrature routines. *The Computer Journal*, 20(2):170–177, 1977.

[65] J. N. Lyness and B. W. Ninham. Numerical quadraure and asymptotic expansions. *Mathematics of Computation*, 21:162–178, 1967.

[66] M. A. Malcolm and R. B. Simpson. Local versus global strategies for adaptive quadrature. *ACM Transactions on Mathematical Software*, 1(2):129–146, 1975.

[67] The Mathworks, Cochituate Place, 24 Prime Park Way, Natick, MA, USA. *MATLAB 6.5 Release Notes*, 2003.

[68] W. M. McKeeman. Algorithm 145: Adaptive numerical integration by Simpson's rule. *Communications of the ACM*, 5(12):604, 1962.

[69] W. M. McKeeman. Algorithm 198: Adaptive integration and multiple integration. *Communications of the ACM*, 6(8):443–444, 1963.

[70] W. M. McKeeman. Certification of algorithm 145: Adaptive numerical integration by Simpson's rule. *Communications of the ACM*, 6(4):167–168, 1963.

[71] W. M. McKeeman and L. Tesler. Algorithm 182: Nonrecursive adaptive integration. *Communications of the ACM*, 6(6):315, 1963.

[72] H. Morrin. Integration subroutine – fixed point. Technical Report 701 Note #28, U.S. Naval Ordnance Test Station, China Lake, California, March 1955.

[73] B. W. Ninham. Generalised functions and divergent integrals. *Numerische Mathematik*, 8:444–457, 1966.

[74] I. Ninomiya. Improvements of adaptive Newton-Cotes quadrature methods. *Journal of Information Processing*, 3(3):162–170, 1980.

[75] S. E. Notaris. Interpolatory quadrature formulae with Chebyshev abscissae. *Journal of Computational and Applied Mathematics*, 133:507–517, 2001.

[76] H. O'Hara and F. J. Smith. Error estimation in Clenshaw-Curtis quadrature formula. *Computer Journal*, 11(2):213–219, 1968.

[77] H. O'Hara and F. J. Smith. The evaluation of definite integrals by interval subdivision. *The Computer Journal*, 12(2):179–182, 1969.

[78] J. Oliver. A practical strategy for the Clenshaw-Curtis quadrature method. *Journal of the Institute of Mathematics and its Applications*, 8:53–56, 1971.

[79] J. Oliver. A doubly-adaptive Clenshaw-Curtis quadrature method. *The Computer Journal*, 15(2):141–147, 1972.

[80] T. N. L. Patterson. Algorithm 468: Algorithm for automatic numerical integration over a finite interval. *Communications of the ACM*, 16(11):694–699, 1973.

[81] J. M. Pérez-Jordá, E. San-Fabián, and F. Moscardó. A simple, reliable and efficient scheme for automatic numerical integration. *Computer Physics Communications*, 70:271–284, 1992.

[82] R. Piessens. An algorithm for automatic integration. *Angewandte Informatik*, 9:399–401, 1973.

[83] R. Piessens, E. de Doncker-Kapenga, C. W. Überhuber, and D. K. Kahaner. *QUADPACK A Subroutine Package for Automatic Integration*. Springer-Verlag, Berlin, 1983.

[84] A. Ralston and P. Rabinowitz. *A first course in Numerical Analysis*. McGraw-Hill Inc., New York, 1978.

[85] J. R. Rice. A metalgorithm for adaptive quadrature. *Journal of the ACM*, 22(1):61–82, 1975.

[86] T. J. Rivlin. *Chebyshev Polynomials: From Approximation Theory to Algebra and Number Theory*. John Wiley & Sons, Inc., New York, Chichester, Brisbane, Toronto, Singapore, 1990.

[87] I. Robinson. A comparison of numerical integration programs. *Journal of Computational and Applied Mathematics*, 5(3):207–223, 1979.

[88] J. H. Rowland and Y. L. Varol. Exit criteria for Simpson's compound rule. *Mathematics of Computation*, 26(119):699–703, 1972.

[89] C. Runge. Über empirische Funktionen und die Interpolation zwischen äquidistanten Ordinaten. *Zeitschrift für Mathematik und Physik*, 46:224–243, 1901.

[90] H. Rutishauser. *Vorlesungen über numerische Mathematik, Band 1*. Birkhäuser Verlag, Basel and Stuttgart, 1976.

[91] H. R. Schwarz. *Numerische Mathematik*. B. G. Teubner, Stuttgart, fourth edition, 1997. With a contribution by Jörg Waldvogel.

[92] H. D. Shapiro. Increasing robustness in global adaptive quadrature through interval selection heuristics. *ACM Transactions on Mathematical Software*, 10(2):117–139, 1984.

[93] R. D. Skeel and J. B. Keiper. *Elementary Numerical Computing with Mathematica*. McGraw-Hill, Inc., New York, 1993.

[94] E. Stiefel. *Einführung in die numerische Mathematik*. B. G. Teubner Verlagsgesellschaft, Stuttgart, 1961.

[95] L. N. Trefethen. Is Gauss quadrature better than Clenshaw–Curtis? *SIAM Review*, 50(1):67–87, 2008.

[96] L. N. Trefethen and J. A. C. Weideman. Two results on polynomial interpolation in equally spaced points. *Journal of Approximation Theory*, 65:247–260, 1991.

[97] A. H. Turetskii. The bounding of polynomials prescribed at equally distributed points (in Russian). *Proc. Pedag. Inst. Vitebsk*, 3:117–127, 1940.

[98] A. Venter. *Aanpasbare Numeriese Integrasie (Adaptive Numerical Integration)*. PhD thesis, Potchefstroom Univeristy for Christian Higher Education, 1995.

[99] A. Venter and D. P. Laurie. A doubly adaptive integration algorithm using stratified rules. *BIT Numerical Mathematics*, 42(1):183–193, 2002.

[100] D. S. Villars. Use of the IBM 701 computer for quantum mechanical calculations II. overlap integral. Technical Report 5257, U.S. Naval Ordnance Test Station, China Lake, California, August 1956.

[101] J. Waldvogel. Fast construction of the Fejér and Clenshaw-Curtis quadrature rules. *BIT Numerical Mathematics*, 46:195–202, 2006.

[102] P. Wynn. On a device for computing the $e_m(s_n)$ transformation. *Mathematical Tables and Other Aids to Computation*, 10(54):91–96, 1956.

Index

Curriculum Vitae

Personal Information

Name	Pedro Gonnet
Birth	January 6, 1978 in Montevideo, Uruguay
Nationalities	Swiss, Canadian and Uruguayan
Languages	English, German French and Spanish

Education

2009	PhD in Computer Science, ETH Zurich
2003	Diploma in Computer Science with a minor (Nebenfach) in Biochemistry, ETH Zurich
1996	Matura Type C, MNG Rämiblil, Zurich

Academic Positions

since 08.2005	Research Assistant in the Group of Prof.W.Gander, Institute of Computational Science, Department of Computer Science, ETH Zurich
02.2002 – 08.2005	Research Assistant in the Group of Prof.P.Koumoutsakos, Institute of Computational Science, Department of Computer Science, ETH Zurich

Work Experience

08.2001 – 02.2002	Researcher at GeneBio S.A.. Geneva, Switzerland
09.1998 – 08.2001	Teaching Assistant (Hilfsassistent) for the Dept. of Computer Science, ETH Zurich
08.1998 – 02.1999	Programmer for GlobalNet Communications, Dbendorf, Switzerland
02.1997 – 08.1997	Programmer for SoftPoint, Montevideo, Uruguay
02.1996 – 02.1997	Layouter for DAZ – Die Andere Zeitung, Zurich, Switzerland

Other Activities

Referee	Journal of Computational Physics; Bioinformatics; Journal of Computational Chemistry; Journal of Bioinformatics and Computational Biology
Teaching Assistant	Informatik I, II, III and IV; Information Systems; Introduction to Computational Science; Computational Science; Software for Numerical Linear Algebra; Parallel Numerical Computing